ORGANIC SPECTROSCOPY

ORGANIC SPECTROSCOPY

Dr. Pradeep Pratap Singh

Assistant Professor
(Department of Chemistry)
Swami Shraddhanand College
University of Delhi
Delhi

Dr. Ambika

Assistant Professor
(Department of Chemistry)
Hansraj College
University of Delhi
Delhi

MV Learning

London • New Delhi

MV Learning
A Viva Books imprint

3, Henrietta Street
London WC2E 8LU
UK

4737/23, Ansari Road,
Daryaganj, New Delhi 110 002
India

ISBN: 978-93-88971-02-7

Printed and bound in India.

Dedicated to our beloved daughter
"EKVIRA"

Preface

The derivation of structural information from spectroscopic data is an integral part of organic chemistry courses at all Universities. At the undergraduate level, the principle aim of courses in organic spectroscopy is to teach students to solve simple structural problems by using combinations of the major techniques (UV, IR, NMR and MS). The courses are taught at the beginning of the third year (6th Semester). By that stage, students are through with an elementary course of organic chemistry (in first year) and a mechanistically-oriented intermediate course (in second year). Students are further exposed, in their honours as well as programme courses, to elementary spectroscopic theory, but are, in general, unable to relate the theory to solve spectroscopic problems.

The book '**Organic Spectroscopy**' deals with UV-Visible, IR, ^{1}H NMR, ^{13}C NMR, and mass spectrometry along with spectroscopic solution of the structural problems. The book covers the basic theory, instrumentation and the structure-spectra correlations of the major spectroscopic techniques. In addition, the book acquaints students with the methods to determine the structure of an unknown compound in a (reasonably) logical manner with the help of all major spectroscopic techniques. Our aim has been to provide spectra to illustrate every point made, but do analyze fully each of the spectra in order to obtain the maximum information available. We believe that learning by solving problems gives more competence and confidence in the subject. So, the book includes, wherever required, numerous solved and unsolved problems that will help students to understand the concepts easily and obtain "structures from spectra". The book is; therefore, a collection of such problems to help students acquaint with the basics of spectroscopy.

We hope our fellow teachers and students for whom this book has been written will find it both interesting and useful. Suggestions for further improvement, along with constructive feedback, from teachers and students, will be acknowledged gratefully. Kindly mail your valuable comments that will help us to improve the book in the next edition.

Pradeep Pratap Singh

 ppsingh@ss.du.ac.in

Ambika

 ambika@hrc.du.ac.in

Contents

Chapter

1

Introduction to Spectroscopy

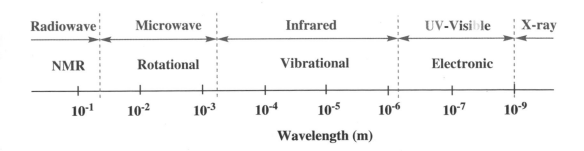

Radiowave	Microwave	Infrared	UV-Visible	X-ray
NMR	Rotational	Vibrational	Electronic	

10^{-1}	10^{-2}	10^{-3}	10^{-4}	10^{-5}	10^{-6}	10^{-7}	10^{-9}

Wavelength (m)

Four techniques are used routinely by organic chemists for structural analysis. Ultraviolet spectroscopy **was the first to come into general use during the 1930s. This was followed by** infrared spectroscopy **in the 1940s, with the establishment of** nuclear magnetic resonance spectroscopy **and** mass spectrometry **during the following two decades. Of these, the first three fall into the category of** absorption spectroscopy.

1 Introduction to Spectroscopy

1.1 SPECTROSCOPY AND SPECTROMETRY

The study of the interaction of energy with matter is known as spectroscopy. When energy is applied to the matter, it can be absorbed, emitted, cause a chemical change or transmitted. This interaction of energy with molecules after interpretation gives information about the molecular structure. The different spectroscopic techniques such as nuclear magnetic resonance (NMR), mass spectrometry (MS), infrared (IR) and ultraviolet-visible (UV-Vis) spectroscopy can be used to determine the structure of the compound.

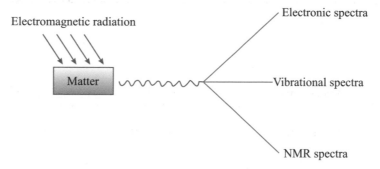

Transitions in all of the other forms of spectroscopy (UV-Vis, IR and NMR) are governed by selection rules state that which transitions are allowed or forbidden? UV-Visible, IR and NMR spectroscopy may be regarded as forms of molecular spectroscopy, while mass spectrometry involves the examination of ions derived from organic molecules and the characteristic fragments arising from their breakdown. Absorption of radiation in the UV-Visible and IR region of electromagnetic spectrum is directly related to the change in the internal energy of the molecules and these in turn are dictated by the molecular structure.

Spectrometry is quite different from usual spectroscopic techniques. It is the method used to acquire a quantitative measurement of the spectrum. It is the practical application where the results are generated. For example, in mass spectrometry the photographic plates are calibrated to measure various values of mass-to-charge ratio (m/z).

1.2 ORIGIN OF MOLECULAR SPECTRA

According to quantum theory of matter the internal energy of a molecule may be raised by the absorption of a quantum of electromagnetic spectrum only if the energy of the quantum exactly equals the difference between the two energy levels in the molecule.

$$\Delta E = hc/\lambda$$

No molecule may absorb energy over the entire electromagnetic spectrum so that in practice absorption is concentrated into certain regions (**Table 1**).

Table 1. Different regions of electromagnetic spectrum.

S. No.	Types of radiations	Range	Types of molecular spectrum
1	Radiowave	> 100 mm	NMR
2	Microwave	1 – 100 mm	Rotational
3	Far IR	50 μm – 1 mm	Vibrational fundamentals or rotational
4	Mid IR	2.5 – 50 μm	Vibrational fundamentals
5	Near IR	780 nm – 2.5 μm	Vibrational (Overtones)
6	UV-Visible	200 – 380 nm	Electronic (Valence orbitals)
		390 – 780 nm	
7	Vacuum UV	10 – 200 nm	Electronic (Valence orbitals)
8	X-rays	10 pm – 10 nm	Electronic (Core orbitals)
9	Gamma rays	10^{-10} cm	Mossbauer
10	Cosmic rays	10^{-12} cm	Mossbauer

1.3 ELECTROMAGNETIC SPECTRUM

The arrangement of all types of electromagnetic radiation in order of their increasing wavelengths or decreasing frequencies is known as electromagnetic spectrum. It ranges from very high frequencies (short wavelength) to very small frequencies (long wavelength). The electromagnetic spectrum is divided into regions discussed below:

1.3.1 Radiowave

The energy of radiowaves is smallest. The nuclear magnetic resonance (NMR) transitions correspond to wavelengths in the radiowave region of the spectrum. It uses the magnetic properties of certain atomic nuclei which relies on the phenomenon of NMR and can provide detailed information about the structure, dynamics, reaction state, and chemical environment of molecules. The intramolecular magnetic field around an atom in a molecule changes the resonance frequency.

1.3.2 Microwave

Microwave (rotational) spectra are very complex for organic molecules and gives only little useful information. Therefore it is rarely used in organic chemistry. Rotational transitions are often responsible for the broadness of infrared (IR) bands, since each vibrational transition has a number of rotational transitions associated with it.

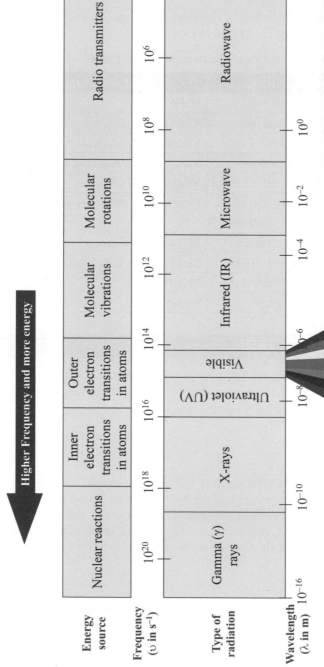

Figure 1. Electromagnetic spectrum.

1.3.3 Infrared

Infrared (IR) spectroscopy deals with the IR region of the electromagnetic spectrum. The absorption of IR is restricted to compounds with small energy differences in the possible vibrational and rotational states. As they involve light with longer wavelength than UV which cannot induce electronic transitions. The IR portion of the electromagnetic spectrum is usually divided into three regions; the near, mid and far IR with respect to the visible spectrum. The higher energy near-IR region ranges approximately $14000 - 4000$ cm^{-1}, mid-IR approximately $4000 - 400$ cm^{-1} and far-IR, approximately $400 - 10$ cm^{-1}.

1.3.4 UV-Visible

Electronic spectroscopy involves the transitions between the electronic levels which have the largest energy gap between transition levels. It uses light in the visible and adjacent (near-UV and near-IR) ranges of electromagnetic spectrum. In this technique absorption measures transitions from the ground state to the excited state. The UV-Vis region is important, since absorptions in this region give rise to the colour associated with molecule. The visible spectrum (between wavelengths from about $390 - 750$ nm) is the small portion of the total electromagnetic spectrum which is visible to the human eye, and therefore it is called visible light.

1.3.5 X-rays

X-rays possess high energy that are capable of ionizing atoms and molecules. They are also important in terms of structure determination using single-crystal X-ray diffraction that provides a "map" of the molecule in the crystal.

1.4 TYPES OF SPECTRA

There are two distinctive classes of spectra: continuous and discrete.

1.4.1 Continuous Spectra

Continuous spectra arise from dense gases or solid objects which radiate their heat away in the form of light. Such objects emit light over a broad range of wavelengths, thus the apparent spectrum seems smooth and continuous. For example: incandescent light bulbs, electric cooking stove burners, flames etc.

1.4.2 Discrete Spectra

Discrete spectra are the observable result of the physics of atoms. With discrete spectra, only bright or dark lines at very distinct and sharply-defined colours are visible. There are two types of discrete spectra, emission and absorption. Discrete spectra with bright lines are called emission spectra and with dark lines are termed as absorption spectra.

1.4.2.1 Emission line spectra

Emission spectrum is produced when the radiant energy is emitted by the excited atom. The excitation of an atom can be carried out by the application of heat, electric charge etc. The electron undergoes excitation from ground state to the excited state and on relaxation it emits the absorbed energy. This emitted radiation is analysed by the spectrometer and the spectra thus obtained is known as emission spectrum. Small changes of energy in an atom generate photons with small energies and long wavelengths, such as radiowaves; whereas large changes of energy in an atom generate high-energy, short-wavelength photons such as UV, X-rays, γ-rays. The emission spectrum for each element is unique as they have their own set of possible energy levels, and with few exceptions the levels are distinct and identifiable. The emission spectra of molecules can be used in chemical analysis of substances.

1.4.2.2 Absorption line spectra

When a substance is subjected to a continuous flow of energy, it absorbs certain wavelength of radiation. The wavelength which is absorbed characterises certain functional groups present in the compound. The dark line which corresponds to the wavelength absorbed is called absorption spectra. An absorption spectrum of a material is the fraction of incident radiation absorbed by it over a range of frequencies. The absorption spectrum is primarily determined by the atomic and molecular composition of the material. Radiation is more likely to be absorbed at frequencies that match the energy difference between two quantum mechanical states of the molecules. The absorption that occurs due to a transition between two states is referred to as an absorption line and a spectrum is composed of many lines.

1.5 DOUBLE BOND EQUIVALENTS (HYDROGEN DEFICIENCY INDEX)

The number of double bond equivalents (DBEs), which tells us how many double bonds or rings, are present in a molecule. Each double bond or ring reduces the number of hydrogens (or halogens) in a molecule by 2, so when we calculate the number of DBEs we simply compare the number of hydrogens which would be present in the fully saturated, acyclic compound with the number actually present, and divide by 2 to give the number of DBEs (**Table 2**).

Table 2. How to calculate the number of double bond equivalents?

Compound	General formula	Double bond / Ring equivalent
Hydrocarbon (C + H)	C_xH_y	$x + 1 - \dfrac{y}{2}$
C + H + divalent atom (O or S)	$C_xH_yO_z$	$x + 1 - \dfrac{y}{2}$
C + H + monovalent atom/Halogen (Cl, Br, I)	$C_xH_yX_z$	$x + 1 - \dfrac{y+z}{2}$
C + H + trivalent atom (N or P)	$C_xH_yN_z$	$x + 1 - \dfrac{y-z}{2}$

Calculating DBE is an important step before analyzing the functional groups and it will offer us more clues to narrow down the options of functional groups and make the work easier.

1.6 DETERMINATION OF STRUCTURE FROM SPECTRA

The first step in the identification of structure of the compound from its spectra is to determine the empirical formula followed by the molecular mass using mass spectrometry. After this the presence of conjugation and the types of functional groups present in the compounds are determined by the use of UV-Visible and IR spectroscopy. Then the number and type of carbon is determined by the ^{13}C NMR. The ^{1}H NMR is generally used to solve the problem at the last stage since it is potentially the most useful for "assembling" the structure. However, at different stage it can be used to support the interpretation (**Table 3**).

Table 3. Systematic use of spectroscopic techniques in the determination of structure.

Spectroscopic technique	Information provided by the technique
Mass Spectrometry (MS)	This technique determines the molecular mass and to identify the presence of isotopes patterns for Cl or Br.
Ultra Violet-Visible Spectroscopy (UV-Visible)	This technique is used to determine the type of chromophores present in a molecule, which chromophore gives rise to the lowest energy transition, whether the system is conjugated or not.
Infrared Spectroscopy (IR)	IR spectroscopy provides the information about the different types of functional groups that are present in the molecule.
^{13}C NMR Spectroscopy	^{13}C NMR spectroscopy provides the information about the carbon skeleton of the molecule, i.e., how many types of carbon and nature of carbon
^{1}H NMR Spectroscopy	^{1}H NMR spectroscopy is probably the last tool to solve the spectra, as: (a) How many types of hydrogen? (b) How many of each type? (c) What types of hydrogen? (d) How are they connected?

UNSOLVED PROBLEMS

1. What do you mean by spectroscopy?
2. What is spectrometry? Give example.
3. Differentiate between spectroscopy and spectrometry?
4. What is an emission spectra?
5. What is an absorption spectra?
6. What is double bond equivalence?
7. What is hydrogen deficiency index?
8. What type of radiations are involved in IR, NMR, and UV spectra?

Chapter

2

UV-Visible Spectroscopy

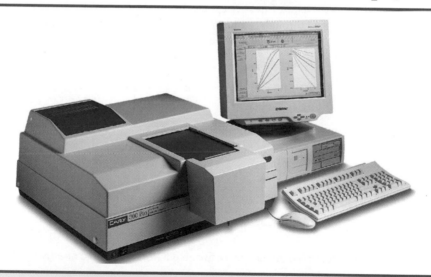

The spectrophotometer was invented in 1940, by Arnold J. Beckman and his colleagues at National Technologies Laboratories, the company Beckman had started in 1935.

UV-Visible Spectroscopy

2.1 INTRODUCTION

Ultraviolet-Visible spectroscopy (electronic spectroscopy) involves light in the visible and adjacent near ultraviolet (NUV) and near infrared (NIR) ranges. It is mostly used for identifying conjugated systems which tend to have stronger absorptions. UV-Visible region of the spectrum is dependent on the electronic structure of the absorbing species like atoms, molecules, ions or complexes. UV-Visible or electronic spectra arise from transitions between electronic energy levels accompanied by changes in both vibrational and rotational states. At room temperature, the majority of the molecules are in the lowest vibrational state of the lowest electronic energy level, (the ground state). When continuous radiation is passed through a transparent material, a portion of it may be absorbed that results in the molecules or atoms to pass from a state of low energy (ground state) to high energy (excited state) **(Figure 1)**.

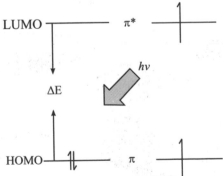

Figure 1. Electronic transitions between different electronic energy levels.

The electromagnetic radiation in the UV-Visible region, absorbed has energy exactly equal to the energy difference between the excited and the ground states. The absorption of energy results in the transitions between different electronic energy levels. As a rule, the portion of energetically favoured electron will occur from the highest occupied molecular orbital (HOMO) to the lowest unoccupied molecular orbital (LUMO), and generates an excited state.

2.2 DIFFERENT REGIONS OF UV-VISIBLE SPECTROSCOPY

The UV-Visible portion of the spectrum is divided into three regions; Vacuum or Far-UV (10 − 200 nm), UV (200 − 380 nm) and Visible (380 − 780 nm) **(Figure 2)**.

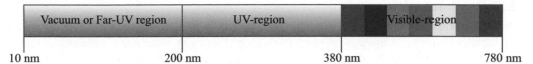

Figure 2. The different regions of UV-Visible spectrum.

The visible region (380 – 780 nm) of the spectrum comprises photons of energies 36 to 72 kcal/mole and the energy range of near ultraviolet region (200 – 380 nm) extends to 143 kcal/mole. The above energies are sufficient to promote or excite a molecular electron to a higher energy orbital.

2.3 DIFFERENT ELECTRONIC TRANSITIONS

The electrons of all the compounds other than alkanes may undergo several possible transitions of different energies (**Figure 3**). On irradiation of the sample molecules with light having an energy that matches a possible electronic transition within the molecule, some of the light energy will be absorbed to promote the electron to a higher energy orbital.

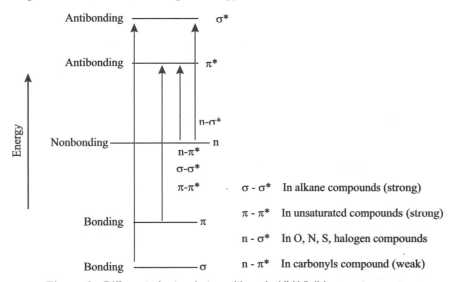

Figure 3. Different electronic transitions in UV-Visible spectroscopy.

For most of the molecules the lowest lying molecular orbitals are the σ bonds. The π-orbitals lie at somewhat higher energy levels, and the nonbonding orbitals, that holds the unshared pair of electrons lie even at high energies. The unoccupied, or antibonding orbitals (σ*, π*), are of highest energy. The usual order of energy required for various electronic transitions is:

$$\sigma \to \sigma^* > n \to \sigma^* > \pi \to \pi^* > n \to \pi^*$$

2.3.1 σ-σ* Transitions

An electron in a bonding orbital (σ) is promoted (excited) to the corresponding antibonding orbital (σ*). Since the σ-bonds are very strong, such transitions (σ-σ*) require large amount of energy (**Figure 3**). For example, methane (alkane) undergoes only σ-σ* transitions and an absorbance maximum appears at 125 nm.

Note: In UV-Visible spectra the absorption maxima due to σ-σ* transitions are not seen, but appears below 200 nm. The study of such transitions are done in vacuum-UV region since below 200 nm the oxygen present in air begins to absorb.

2.3.2 n-σ* Transitions

In this transition an electron in a non-bonding orbital (n) is excited to the antibonding orbital (σ*) and generally require less energy than σ-σ* transitions **(Figure 3)**. Saturated compounds containing one heteroatom with non-bonding electrons undergoes n-σ* transitions, such as alcohols, ethers, amines and sulfur compounds. For example, in saturated alkyl halides, the energy required for this transition decreases with an increase in the size of the halogen atom (or a decrease in the electronegativity of the atom). Due to the greater electronegativity of the chlorine atom (than iodine) the non-bonding (n) electrons on chlorine are comparatively difficult to excite, as these electrons are tightly bounded to the nucleus. Since, this transition is more probable in methyl iodide so its molar excitation coefficient is also higher than methyl chloride. These transitions are sensitive to hydrogen bonding. For example, alcohols form hydrogen bonds with the solvent molecules which occurs due to the presence of non bonding electrons on the heteroatom and thus, n-σ* transitions require greater energy.

2.3.3 π-π* Transitions

These transitions take place in molecules containing an unsaturated group to provide π-electrons, for example, alkenes, alkynes, aromatic compounds, carbonyl compounds etc., and require less energy than n-σ* transitions **(Figure 3)**. In the case of alkenes, there are several transitions possible, but the π-π* are of lowest energy and are observed nearly 175 – 190 nm. The saturated carbonyl compounds show the π-π* transitions (~150 nm) in addition to n-π*.

2.3.4 n-π* Transitions

Compounds containing one heteroatom with non-bonding electrons are capable of n-π* transitions. These transitions require the least amount of energy than all other transitions and therefore absorption bands are generally observed at longer wavelengths **(Figure 3)**. Carbonyl-unsaturated systems incorporating nitrogen or oxygen atoms can undergo n-π* transitions (~285 nm). Despite of the fact this transition is forbidden by the selection rules, it is the most observed and studied transition for carbonyls. This transition is also sensitive to constituents on the carbonyl.

2.4 SELECTION RULE

There are many electronic transitions possible, but it does not mean that they can or will occur. There are complex selection rules based on the symmetry of the ground and excited states of the molecules under examination. The following selection rules are operative.

  (i) Electronic transitions are allowed if the orientation of the electron spin does not change during the transition.
  (ii) If the symmetry of the final and initial functions is different these are called the *spin and symmetry selection rules*, respectively. However, the forbidden transitions can still occur, but will give rise to weak absorptions.

2.5 ORIGIN OF UV-VISIBLE SPECTRUM

UV-Visible absorption bands are characteristically broad, although energy transitions between vibrational or rotational energy levels within one electronic level can show fine structures. This can be attributed to the fact that the electronic transitions are accompanied by vibrational and rotational transitions, so that the promotion of an electron can occur from the ground state electronic energy level to any vibrational or rotational energy levels **(Figure 3)**. When the sample is in gaseous or vapour phase, the spectrum consists of a number of closely spaced lines, constituting a band spectrum.

However, in the solution phase, the absorbing species are surrounded by solvent molecules and due to solvent-solute interaction; the spectrum acquires the shape of a smooth and continuous absorption peak.

EXERCISE PROBLEMS 2.1

1. What types of electronic transitions are possible for each of the following compounds:
   (i) Acetone; (ii) Cyclopentene; (iii) Acetaldehyde; (iv) Dimethyl ether; (v) Methyl vinyl ether; (vi) Trimethylamine; (vii) Cyclohexane; (viii) Methanol; (ix) Carbon tetrachloride; (x) Methyl chloride; (xi) Ethyl chloride; (xii) Formaldehyde; (xiii) Vinyl chloride.
2. Explain, why methyl chloride absorbs at 173 nm, while methyl iodide absorbs at 279 nm in hexane.
3. Explain, why a polar solvent usually shifts the n-π^* transition to a shorter wavelength and π-π^* transition to a longer wavelength.

2.6 DESIGNATION OF UV-BANDS

The absorption bands in the UV-Visible spectrum may be designated either by using electronic transitions or by the letter designation.

2.6.1 R-Bands (German, Radical like)

The R-bands are formed due to n-π^* transitions of single chromophoric group, such as, the carbonyl or nitro group. They are characterised by low molar absorptivity (ε_{max} < 100) and undergo hypsochromic shift with an increase in the solvent polarity.

2.6.2 K-Bands (German or Conjugated)

The K-band arises due to π-π^* transitions in the molecules containing conjugated π-systems, for example, butadiene, mesityl oxide etc. They are characterised by low molar absorptivity (ε_{max} > 10,000).

2.6.3 B (Benzenoid) and E (Ethylenic) Bands

The spectra of aromatic or hetero-aromatic compounds exhibit E and B bands representing π-π^* transitions. For example, E_1 and E_2 bands of benzene occurs near 180 and 200 nm and their molar absorptivity varies between 2000 and 14000. The B-band occurs in the region from 250 – 255 nm as a broad band containing multiple fine structures and represents a symmetry-forbidden transition, which has finite but low probability due to forbidden transition in the highly symmetrical benzene molecule. The vibrational fine structure appears only in the B-band and disappears frequently in the more polar solvents.

2.7 BEER-LAMBERT LAW

According to the Beer-Lambert law, the absorbance A of the solution is directly proportional to the path length (l = the length of the cell containing the solution, in cm) and the concentration of the absorbing species (c, in moles per litre).

$$A = \log(I_0/I) = \varepsilon.l.c$$

Where, A is the measured absorbance, I_0 is the intensity of the incident light at a given wavelength, I is the transmitted intensity, ε is a constant known as the molar absorptivity or molar extinction coefficient, and is a characteristic of the molecule. It is a measure of the intensity of the absorption and usually ranges from 0–10^6 (unit of $100 \text{ cm}^2 \text{ mol}^{-1}$). The greater the probability of the absorption and its associated electronic transitions, greater the ε value for that transition. For most molecules, absorptions associated with π-π^* transitions have lower ε values than n-σ^* transitions. In general, forbidden transitions give rise to low intensity (low ε) absorption bands ($\varepsilon < 10,000$), but two important "forbidden" absorption are commonly seen: n-π^* transitions of ketones at approximately 300 nm (ε usually $10 - 100$), and the weak π-π^* absorption of benzene rings at 260 nm (ε about $100 - 1000$) approximately.

Figure 4. UV-Visible spectrophotometer.

2.8 INSTRUMENTATION FOR UV-VISIBLE SPECTROSCOPY

An optical spectrometer records the wavelength at which absorption occurs, together with the degree of absorption at each wavelength. The resulting spectrum is presented as a graph of absorbance (A) versus wavelength (λ). Absorbance usually ranges from 0 (no absorption) to 2 (99% absorption), and is precisely defined in context with spectrometer operation (**Figure 4**).

2.8.1 Single Beam Spectrometers

Single beam spectrometer consists of a single beam of light which is used for reading the absorption of the sample as well as the reference. The radiation from the source is passed through a filter or a suitable monochromator to get a monochromatic radiation. Then, it is passed through the sample (or the reference), the transmitted radiation is detected by the photo-detector and then is recorded (**Figure 5**).

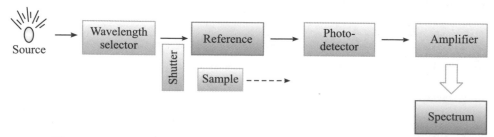

Figure 5. Schematic diagram for a single beam UV-Visible spectrometer.

2.8.2 Double Beam Spectrometers

In a double beam spectrometer, the radiation from the monochromator splits up into two beams with the help of a beam splitter and is passed simultaneously through the reference and the sample cell. The transmitted radiations are detected by the detectors and the difference in the signal at all the wavelengths is suitably amplified and recorded (**Figure 6**).

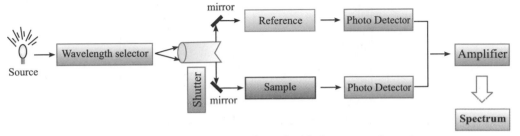

Figure 6. Schematic diagram for a double beam spectrometer.

2.8.3 Photodiode Array Spectrometer (Multi-channel Instrument)

In a photodiode array instrument, the radiation from the source is focused directly on the sample. Thus, the radiations of all the wavelengths fall simultaneously on the sample. The radiation coming out of the sample after absorption (if any) is then made to fall on a reflection grating (**Figure 7**).

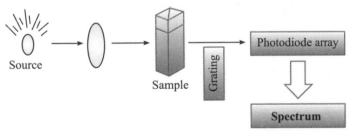

Figure 7. Schematic diagram for a photodiode array spectrometer.

The grating disperses all the wavelengths simultaneously. Then, these fall on the array of the photodiodes arranged side by side. In this way the intensities of all the radiations in the range of the spectrum are measured in one go. The advantage of such instrument is that a scan of the whole range can be accomplished in a short time.

2.9 SOME USEFUL TERMS IN ULTRAVIOLET-VISIBLE SPECTROSCOPY

2.9.1 Solvents used in the UV-Visible Spectrophotometry

The solvent plays an important role in the UV-Visible spectroscopy, therefore, a good solvent should not absorb UV radiation in the same region as the substance whose spectrum is to be determined. Also, solvent has large effect on the fine structure of an absorption band. For example, the non polar solvent does not H-bond with solute, and the spectrum of the solute closely approximates the spectrum that would be produced in the gaseous state, where fine structures can often be observed. In polar solvents, the hydrogen bonding forms a solute-solvent complex, and the fine structure may disappear (**Figure 8**).

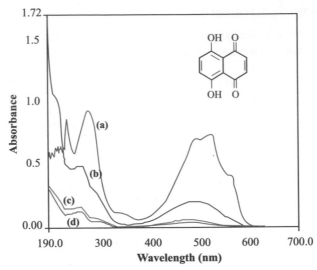

Figure 8. Effect of solvent on the UV-Visible spectra of 9,10-dihydroxynapthoquinone (a) Hexane; (b) $CHCl_3$; (c) MeOH; (d) H_2O.

Table 1. Wavelength (λ_{max}) values of some common solvents used in UV-Visible spectrophotometry.

Solvents	λ_{max} values	Solvents	λ_{max} values
Acetonitrile	190	n-Hexane	201
Chloroform	240	Methanol	205
Cyclohexane	195	Isooctane	195
1,4-Dioxane	215	Water	190
95% Ethanol	205		

The solvents also influence the wavelength of absorption peaks. For example, polar solvents shift transitions of the n-π* type to shorter wavelength. Water, ethanol and hexane are commonly used solvents as they remain transparent in the UV-region of the spectrum, where, interesting absorption peaks from the sample molecules are likely to occur (**Table 1**).

2.9.2 Chromophores

The colour of the coloured substances is due to the presence of one or more unsaturated groups responsible for electronic absorption. These groups are called chromophore, e.g., C=C, C≡C, C=N, C≡N, C=O, N=N, NO_2 etc.

Table 2. Absorptions of simple chromophores in the UV-Visible spectrophotometry.

Chromophore	Transitions	λ_{max} (nm)	Examples
Alcohol (ROH)	n - σ*	180	Methyl alcohol
Alkene (C=C)	π - π*	171	Ethylene
Alkyne (C≡C)	π - π*	170	Acetylene
Alkyl halide (X), X = Cl	n - σ*	173	Methyl chloride
X = Br	n - σ*	208	n-Propyl bromide
X = I	n - σ*	259	Methyl iodide

Chromophore	Transitions	λ_{max} (nm)	Examples
Aldehyde (CHO)	n - π* π - π*	290 180	Acetaldehyde
Ketone (C=O)	n - π* π - π*	280 180	Acetone
Acid (COOH)	n - π*	205	Acetic acid
Ester (COOR)	n - π*	205	Ethyl acetate
Amide (CONH$_2$)	n - π*	210	Acetamide
Nitro (NO$_2$)	n - π* π - π*	275 200	Nitrobenzene
Amine (NH$_2$)	n - σ*	190	Trimethylamine
Nitrile (C≡N)	n - π*	160	Acetonitrile

When, sample molecules are exposed to UV-light having an energy that matches a possible electronic transition within the molecule, some of the light energy will be absorbed as the electron is promoted to a higher energy orbital. The nuclei that hold the electrons together in bonds play an important role in determining which wavelengths of radiation are absorbed. The nuclei determine the strength with which the electrons are bound and thus influence the energy spacing between ground and excited states. Hence, the characteristic energy of a transition and the wavelength of the radiations absorbed are properties of a group of atoms rather than of electrons themselves. Since similar functional groups will have electrons capable of discrete classes of transitions, the characteristic energy of these transitions is more representative of the functional group than the electrons themselves. Structural or electronic changes in the chromophore can be quantified and used to predict shifts in the observed electronic transitions (**Table 2**).

2.9.3 Effect of Substitution on the Chromophore

Substituents may have the following effects on a chromophore.

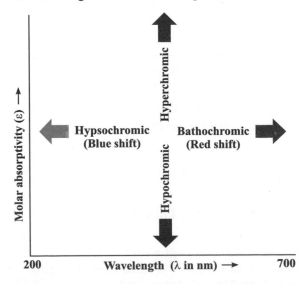

Figure 9. Effect of substituents on the position and intensity of an absorption band.

2.9.3.1 Bathochromic shift (Red shift)

The shift of an absorption maximum towards longer wavelength (lower energy) is commonly known as *bathochromic shift* or *red shift* (**Figure 9**). It may be produced by a change of medium (solvent) or by the presence of an auxochrome.

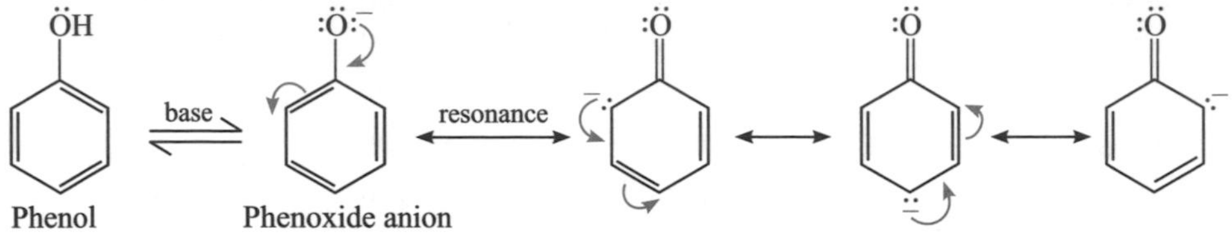

Phenol
1

Phenoxide anion
1a

Scheme 1

For example, phenol (**1**) and substituted phenol are acidic and display striking changes to their absorptions upon the addition of base by the formation of phenoxide ion (**1a**). The removal of the acidic proton increases the conjugation of the lone pairs on the oxygen with the π-system of the aromatic ring, leading to a decrease in the energy difference between the HOMO and LUMO orbitals, resulting in red shift, along with an increased intensity of the absorption (**Scheme 1**).

2.9.3.2 Hypsochromic shift (Blue shift)

The shift towards shorter wavelength (higher energy) is known as *hypsochromic shift* or *blue shift* (**Figure 9**). This may be caused by the change of medium (solvent) and also by the removal of conjugation. For example, when aromatic amine is treated with acid, a blue shift is observed with a decrease in intensity which is due to the loss of the conjugation of the lone pair of electrons on the nitrogen atom of aniline (**2**) with the π-bond system of the benzene on protonation to form anilinium ion (**2a**) (**Scheme 2**).

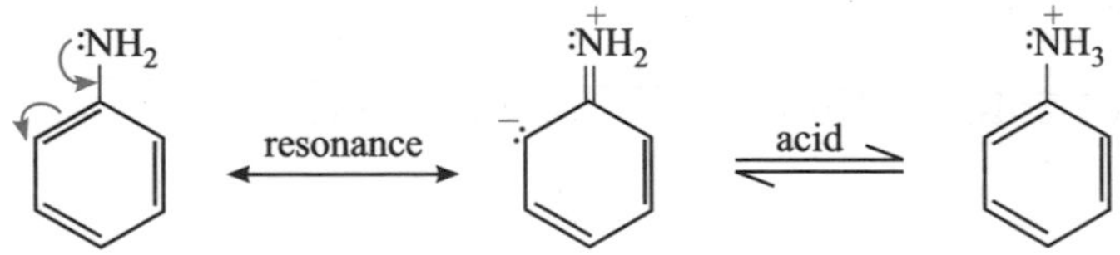

Aniline
(Neutral or basic condition)
2

Protonated
(Acidic condition)
2a

Scheme 2

2.9.3.3 Hyperchromic shift

An increase in intensity leads to the *hyperchromic shift* (**Figure 9**). For example, the intensities of primary and secondary bands of phenol are increased in phenolate.

2.9.3.4 Hypochromic shift

A decrease in intensity leads to the *hypochromic shift* (**Figure 9**). For example, the intensities of primary and secondary bands of benzoic acid are decreased in benzoate.

2.9.4 Effect of Conjugation on the Chromophore

The conjugation of double and triple bonds shift the absorption maximum to longer wavelength. Each additional double bond in the conjugated π-electron system shifts the absorption maximum about 30 nm in the same direction. Also, with each new conjugated double bond the molar absorptivity (ε) roughly doubles. Thus, extending conjugation generally results in *bathochromic* and *hyperchromic shifts* in absorption. This can be explained on the basis of the relative energy levels of the π-orbitals. When two double bonds are conjugated, the four π-atomic orbitals combine to generate four π-molecular orbitals (two are bonding and two are antibonding). The energetically most favorable π-π* transitions occur from the highest energy bonding π-orbital (HOMO) to the lowest energy antibonding π-orbital (LUMO). Increased conjugation brings the HOMO and LUMO orbitals closer together. Therefore, the energy (ΔE) required for the promotion of electron is less, and hence the wavelength increases correspondingly ($\lambda = hc/\Delta E$). Thus, *more highly conjugated the system, smaller the* HOMO-LUMO *gap, E, and therefore lower the frequency and longer the wavelength.*

2.9.5 Effect of Solvent on the Chromophore

The solvent in which the absorbing species is dissolved also has an effect on the spectrum of the species. According to the Frank Condon principle the electronic transitions involve the movement of electrons, including those of the solvents, but not the movement of atoms. When the solvent electrons rearrange to stabilise the excited state of the molecule, the energy difference between electronic levels of the molecules is lowered and the absorption moves to higher wavelength. For example, the peaks resulting from n-π* transitions of carbonyl compounds (ketones) get shifted to shorter wavelength (*blue shift*) with increasing solvent polarity due to the increased solvation of the lone pair, which lowers the energy of the n orbital and thereby increasing the energy required to promote an electron to π* energy level. Often the *red shift* is observed for the π-π* absorber, which lower the energy levels of both the excited and unexcited states. This effect is greater for the excited state, and so the energy difference between the excited and unexcited states is slightly reduced, resulting in a small *red shift*. This effect also influences n-π* transitions but is over shadowed by the *blue shift* resulting from solvation of lone pairs.

2.9.6 Effect of pH on the Chromophore

The UV-Visible spectra of some compounds exhibit drastic changes with a change in the pH of the solvent. The utility of many pH indicators is due to their absorption in the visible region of the UV-Visible spectrum. The change in the pH changes results in the change in indicator chromophore, that leads to a reliable colour change at predictable pH values. For example, phenolphthalein (**3**) can be deprotonated at elevated pH to give the corresponding anions (**3a**), resulting into the extended conjugation of the chromophore that leads to *bathochromic shift* (**Scheme 3**). Thus, the anion of phenolphthalein is deep magenta in colour, while un-ionised form of phenolphthalein is colourless.

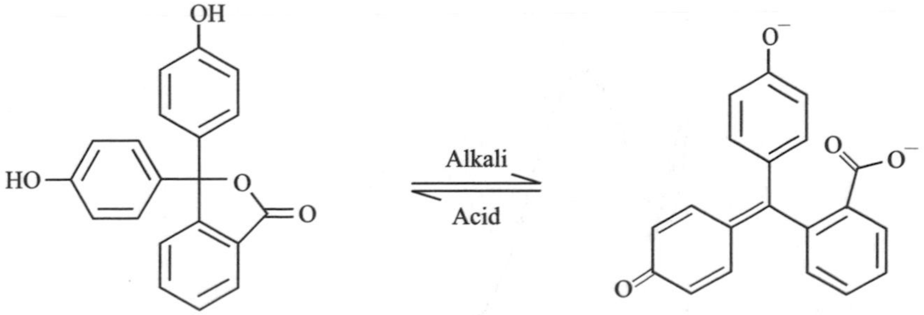

Scheme 3

The pK_a of the acid-base equilibrium is 9.4. At acidic and neutral pH there is insufficient anions to detect the colour by eye, and it appears colourless. As the pH approaches the pK_a, the concentration of anion increases and at pH 8.2, the colour becomes visible to the eye. As the pH 8.2 is close to neutrality, phenolphthalein is widely used to show the end point in weak acid-strong base titrations.

2.9.7 Effect of Steric Hindrance on the Chromophore

The UV-Visible spectroscopy is very sensitive to distortion of the chromophore. Thus, the steric repulsion which opposes the coplanarity of conjugated electron systems can be easily detected by comparing its UV-spectrum with that of a model compound. Distortion of the chromophore may lead to *red* or *blue shifts* depending upon the nature of the distortion. For example, biphenyl **(4)** and its 2-methyl **(5)** and 2,2'-dimethyl **(6)** analogs, show steric inhibition of resonance. Biphenyl is not exactly coplanar but is twisted by a very small angle. Thus, conjugation between the rings is not affected. Addition of bulky substituents in the ortho positions lead to more twisting and loss of conjugation. Due to the effect of forcing the ring out of coplanarity in 2,2'-dimethylbiphenyl **(6)** the absorption patterns are similar to those of *o*-xylene **(Figure 10)**.

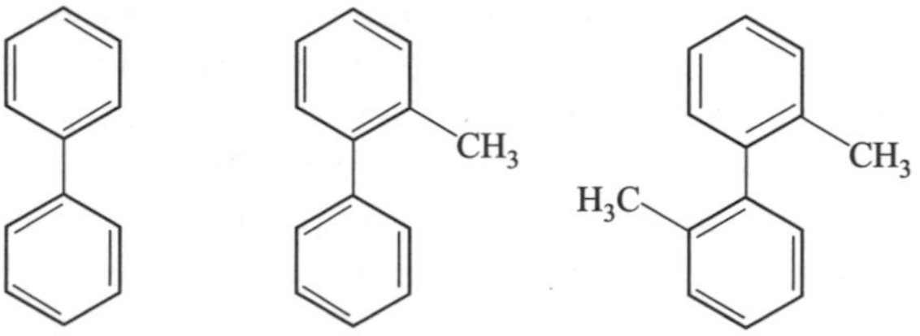

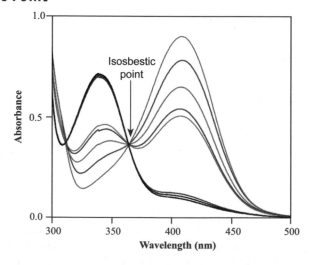

Figure 10. Structure of biphenyl (4), 2-Methylbiphenyl (5) and 2,2'-Dimethylbiphenyl (6).

2.9.8 Auxochromes

An auxochrome is a group which interacts with the chromophore and enhances its colour imparting properties without being itself a chromophore causing a *bathochromic shift*. The most important property of an auxochrome is its ability to provide additional opportunity for charge delocalisation and thus providing smaller energy increments for transition to excited states. For example, hydroxyl, alkoxy, amino and halogen groups.

2.9.9 Isosbestic Point

Figure 11. UV-Visible spectra indicating the isosbestic point.

The presence of the two species in equilibrium can be identified by the appearance of an *isosbestic point* in the UV spectrum. If two substances, each of which obeys Beer's law, are in equilibrium, the spectra of all the equilibrium mixtures at a total concentration intersect at a fixed wavelength. This point is the wavelength at which absorbance of two species are equal and is termed as *isosbestic point* (**Figure 11**).

EXERCISE PROBLEMS 2.2

1. Explain, why benzene shows an absorption band at λ_{max} 254 nm (ε_{max} = 230) while aniline absorbs at λ_{max} = 280 nm?

2. Why aniline shows absorption bands at λ_{max} 280 and 230 nm while in aqueous acidic medium it absorbs at λ_{max} – 254 and 203 nm.

3. Why phenol absorbs at λ_{max} 270 and 210 nm in aqueous solution, while in alkaline medium it absorbs at λ_{max} = 287 and 235 nm.

4. Biphenyl exhibits a very intense absorption band at 252 nm, while 2,2′-dimethyl derivative spectrum has an absorbtion band of low intensity at 270 nm. Explain.

2.10 WOODWARD-FIESER RULES

The absorption maximum of the highly substituted diene shifts to the longer wavelength by about 15 nm in the UV-Visible spectrum. This "substituent effect" is general for dienes, trienes, and is even more pronounced for enone chromophores. Woodward and Fieser performed extensive studies of terpene and steroidal alkenes, and noted similar substituents and structural features would predictably lead to an empirical prediction of the wavelength for the lowest energy π-π* electronic transition. The Woodward-Fieser rules have been tabulated in table 3, 4 and 5 using these rule the λ_{max} of the different diene, enones and aromatic carbonyl compounds can be calculated.

2.10.1 Woodward-Fieser Rules for Calculating the λ_{max} Values in Dienes

The conjugated diene exhibits an intense band in the region 217 to 245 nm, due to the π-π* transition. The position of this band is insensitive to the nature of solvent. Many simple conjugated dienes exists in a planar conformation. Generally, alkyl substituents produce *bathochromic* and *hyperchromic shifts*.

However, with certain patterns of alkyl substitution, the wavelength increases and the intensity decreases. The 1,3-dialkylbutadienes possess too much crowding between alkyl groups to permit them to exist in the *s-trans* conformation. They convert by rotation around the single bond to *s-cis* conformation which absorbs at longer wavelength but lower intensity than the corresponding *trans*-conformation. But in cyclic dienes, where the central bond is a part of the ring system, the diene chromophore is held rigidly in either the *s-trans* (transoid) or the *s-cis* (cisoid) orientation. The incremental contribution of substituents is added to this base value from (**Table 3**).

Table 3. Empirical rules for calculating λ_{max} values in case of dienes.

Diene with appropriate parent value	Add an increment for any extra conjugated π-bonds in specific dienes	Add increments to any substituents in specific dienes	
(a) Acyclic (7a) and heteroannular (7b) diene (double bonds in two adjacent ring) = 217 nm R~~~~~~R **7a** **7b**	C=C (Exocyclic double bond) **8a** +5 C=C (Double bond extending conjugation) **8b** +30	R (CH$_3$) Cl OH, OR OCOR O$^-$ SR NR	+5 +5 +5 0 – +30 +60
(b) Homoannular diene (double bonds within one ring) = 253 nm **7c**			

2.10.1.1 Acyclic dienes

The base value for the acyclic dienes is 217 nm.

Examples to calculate the λ_{max} of different acyclic dienes:		
1.	Acyclic butadiene	= 217 nm
	One alkyl substituent	= +5 nm
	Calculated value	= 222 nm
	Experimental value	= 220 nm
2.	Acyclic butadiene	= 217 nm
	2 Alkyl substituent (2 × 5)	= +10 nm
	1 Exocyclic C=C	= +5 nm
	Calculated value	= 232 nm
	Experimental value	= 237 nm

2.10.1.2 Cyclic dienes

There are two types of cyclic dienes, with two different base values, i.e., *heteroannular* (7b, 217 nm) and *homoannular* (7c, 253 nm) dienes.

Examples to calculate the λ_{max} of different cyclic dienes:

1.

Exocyclic C=C

Heteroannular diene	= 217 nm
3 Alkyl substituents (3 × 5)	= +15 nm
1 Exocyclic C=C	= + 5 nm
Calculated value	= 237 nm
Experimental value	= 235 nm

2.

Exocyclic C=C

Heteroannular diene	= 217 nm
4 Alkyl substituents (4 × 5)	= +20 nm
1 Exocyclic C=C	= + 5 nm
Calculated value	= 242 nm
Experimental value	= 240 nm

3.

Exocyclic C=C

Homoannular diene	= 253 nm
4 Alkyl substituents (4 × 5)	= +20 nm
1 Exocyclic C=C	= + 5 nm
Calculated value	= 278 nm
Experimental value	= 275 nm

2.10.2 Woodward-Fieser Rules for Calculating the λ_{max} Values in Unsaturated Carbonyl Compounds (Enones)

The carbonyl compounds have two main UV-transitions (i) π-π^*, (ii) n-π^*. The conjugation of the double bond with a carbonyl group leads to the intense absorption corresponding to π-π^* transitions of the carbonyl group. The absorption in simple enones normally occurs in the region between 220 and 250 nm. The n-π^* transition is less intense and occurs in between 310–330 nm. The structural modifications of the chromophores have pronounced effect on the π-π^* transitions, but no effect on the n-π^* transitions. The incremental contribution of substituents is added to this base value from the (**Table 4**).

Table 4. Empirical rules for calculating λ_{max} values in case of different enones.

Enones $\overset{\beta\quad\alpha\quad O}{\underset{}{—C=C—C—X}}$	Enones with parent value	Add an increment for any extra conjugated π-bond in specific 'Enones'		Add increments to any substituent in specific 'Enones'				
				R	α	β	γ	δ
If, X = H	207 nm	C=C (Exocyclic double bond)	+5	CH$_3$	10	12	18	18
X = C (part of five membered ring)	202 nm			Cl	15	12	–	–
		C=C (Double bond extending conjugation)	+30	OH	35	30	–	–
				OR	6	6	–	–
		Homoannular diene	+39	OCOR	–	75	–	–
X = C (Acyclic or part of six-membered ring)	215 nm			SR	–	95	–	–

Examples to calculate the λ_{max} of different enones:		
1.	Acyclic enone	= 215 nm
	α-Alkyl substituents	= +10 nm
	β-Alkyl substituents	= +12 nm
	Calculated value	= 237 nm
	Experimental value	= 240 nm
2.	Cyclic six memberd enone	= 215 nm
	C=C extending conjugation	= +30 nm
	Homoannular diene	= +39 nm
	δ-Ring residue	= +18 nm
	Calculated value	= 302 nm
	Experimental value	= 300 nm
3.	Cyclic five memberd enone	= 202 nm
	OH α-substituent	= +35 nm
	β-ring residue	= +12 nm
	Calculated value	= 249 nm
	Experimental value	= 249 nm
4.	Cyclic six memberd enone	= 215 nm
	C=C extending conjugation	= +30 nm
	γ-Ring residue	= +18 nm
	δ-Ring residue	= +18 nm
	Exocyclic double bond	= +5 nm
	Calculated value	= 286 nm
	Experimental value	= 288 nm
5.	Cyclic six memberd enone	= 215 nm
	C=C extending conjugation	= +30 nm
	Homoannular diene	= +39 nm
	γ-Ring residue	= +18 nm
	δ-Ring residue (2 × 18)	= +36 nm
	Exocyclic double bond	= +5 nm
	Calculated value	= 343 nm
	Experimental value	= 339 nm

2.10.3 Woodward-Fieser Rule for Calculating the λ_{max} Values in Aromatic Carbonyl Compounds

Woodward-Fieser rule can be used to calculated the λ_{max} value for the π-π* transitions (**Table 5**).

Table 5. Empirical rules for calculating λ_{max} values in case of different aromatic carbonyls.

Aromatic carbonyls	Aromatic carbonyls with appropriate parent value	Add increment to any substituent in specific Enones		
		o	*m*	*p*
If, X = H	250 nm	CH_3	3	3	7
X = C	246 nm	Cl	0	0	10
X = OH, OR	230 nm	OH, OR	7	7	25
		OCOR	–	–	–
		O^-	15	15	80
		SR	–	–	–
		NR_2	20	20	85

Depending on the nature of aromatic carbonyl compounds different base values exists. For aromatic aldehydes, aromatic ketones and aromatic carboxylic acids or esters, base values are 250 nm, 246 nm and 230 nm respectively.

Examples to calculate the λ_{max} of different aromatic carbonyl compounds:		
1.	Aromatic carbonyl compound	= 246 nm
	–Cl substitution at para position	= +10 nm
	Calculated value	= 256 nm
	Experimental value	= 254 nm
2.	Aromatic carbonyl compound	= 246 nm
	– OH substitution at meta position	= +7 nm
	– OH substitution at para position	= +25 nm
	Calculated value	= 278 nm
	Experimental value	= 281 nm
3.	Aromatic carbonyl compound	= 230 nm
	–Br substitution at para position	= +15 nm
	Calculated value	= 245 nm
	Experimental value	= 245 nm

EXERCISE PROBLEMS 2.3

1. 1,3-butadiene absorbs at 217 nm, but 1,5-hexadiene absorbs below 200 nm. Explain.

2. Explain, why butadiene has a higher λ_{max} value (π-π* transition) than ethylene.

3. Predict whether UV-Vis spectroscopy can be used to distinguish between the following isomers. Estimate λ_{max} (there may be more than one) for each.

(a) $CH_2=CH-CH_2-CH=CH-CH_3$ and $CH_3-CH=CH-CH=CH-CH_3$

(b) $CH_3-CO-O-CH_2-CH_3$ and $CH_3-CH_2-CO-O-CH_3$

(c) ⬠—CH_2CN   and   ⬠=CH–CN

(d) Homoannular and heteroannular dienes

(e) Ethyl benzene and styrene

(f) *cis-* and *trans-* stilbene

2.11 APPLICATIONS OF UV-VISIBLE SPECTROSCOPY

2.11.1 Detection of Functional Groups

The UV-Visible spectroscopy can be used to identify the presence of a chromophore. The absence of a band at a particular wavelength may be regarded as an evidence for the absence of a particular group in the compound. If the spectrum of the compound is transparent above 200 nm, it indicates the absence of conjugation, carbonyl group, aromatic compound and bromo or iodo atoms. An isolated double bond or some other atoms or groups may be present.

2.11.2 Extent of Conjugation

The extent of conjugation can be estimated by using this spectroscopic technique. Addition in unsaturation with the increase in the number of double bonds, shifts the absorption to longer wavelength.

2.11.3 Distinction between Conjugated and Non-conjugated Compounds

It also distinguishes between conjugated and non-conjugated compounds. The forbidden band for the carbonyl group in the compound **9** will appear at longer wavelength than that of the compound **10**.

9   10

2.11.4 Identification of an Unknown Compound

An unknown compound can be identified by comparing its spectrum with the known spectra. If the two spectra coincide, the two compounds must be identical, if not, then the expected structure is different from the known compound.

2.11.5 Quantitative Analysis

UV-Visible spectroscopy can be used to determine the concentration of a particular compound in a solution as it is based on the Beer-Lambert law (see section 2.7). Thus, for a fixed path length it is necessary to know how quickly the absorbance changes with concentration.

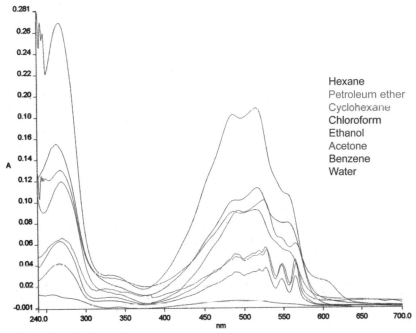

Figure 12. UV-Visible spectra of different solvent extracts of Arnebia species.

For example, the extraction of coloured pigments from Arnebia species with different solvents: n-Hexane, cyclohexane, $CHCl_3$, ethanol, dichloromethane, and water for 24 hours at room temperature can be calculated using UV-Visible spectroscopy. The spectrum can be recorded to measure the absorbance from Lambert-Beer's law of the standard solution of the extracts. Thus, the concentration of the colouring matter extracted can be calculated in different solvents (**Figure 12**).

2.11.6 Examination of Polynuclear Hydrocarbons

Benzene and polynuclear hydrocarbons have characteristic spectra in the UV-Visible region. Thus, the identification of the polynuclear hydrocarbons can be made by comparison with the spectra of known polynuclear compounds. The presence of substituents on the ring generally shifts the absorption maximum to longer wavelength. For example, the naphthalene shows absorption at 210 and 272 nm, the added conjugation in anthracene and tetracene causes bathochromic shift of these absorption bands (**Figure 13**).

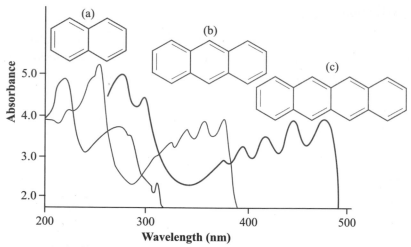

Figure 13. UV-Visible spectra of polynuclear hydrocarbons: (a) Naphthalene; (b) Anthracene; (c) Tetracene.

2.11.7 Determination of the Strength of Hydrogen Bonding

Hydrogen bonding involves electron cloud transfer from H-atom to the neighbouring electronegative atom and occurs in any system containing a proton donor group and a proton acceptor provided the *s*-orbital of the proton can effectively overlap the p or π-orbital of the acceptor group. The formation of the H-bonding between molecule of solvent and the solute can have a profound effect on the spectral characteristics of the solute. Therefore, change in the solvent polarity may result in a change in the excitation energy of the molecule. For example, acetone dissolved in hexane absorbs at 279 nm but at 264 nm when dissolved in water. This could be attributed to the fact that when a carbonyl compound such as acetone is dissolved in polar solvent such as water, H-bonds are formed between the solvent and non-bonding (n) electrons of the carbonyl oxygen. This lowers the energy of the n-orbitals in the ground state by an amount equivalent to the strength of H-bond.

2.11.8 Study of Charge Transfer Complexes

The formation of the charge transfer complexes occurs between molecules which, when mixed, allow transfer of the electronic charge through space from an electron rich molecule to an electron deficient molecule.

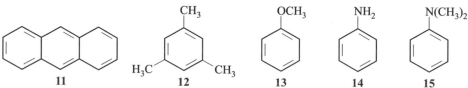

Figure 14. Charge transfer donors used in the study of charge transfer complexes.

The common donors and acceptors are given in **figure 14** and **15**.

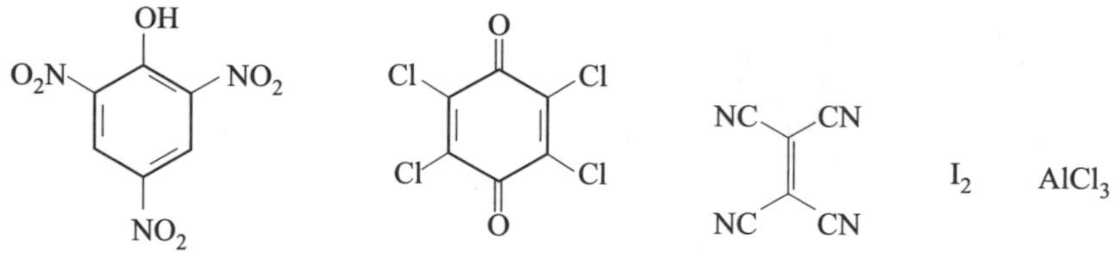

16 **17** **18** **19** **20**

Figure 15. Charge transfer acceptors used in the study of charge transfer complexes.

The bond formation between molecules occurs when filled π-orbitals in the donor overlap with the depleted orbitals in the acceptors resulting in the production of two new molecular orbitals. The transitions between the newly formed orbitals are responsible for the new absorption bands observed in the charge transfer complexes. According to the charge transfer spectra, electron transfer from the donor to the acceptor is more complete in the excited state than in the ground state and the wavelength of absorption can be correlated with electron affinity of the acceptor and the ionisation potential of the donors. For example, iodine imparts violet colour in benzene, while it is brown in hexane. This is due to the formation of the charge transfer complex between the pairs of molecules (**Scheme 4**).

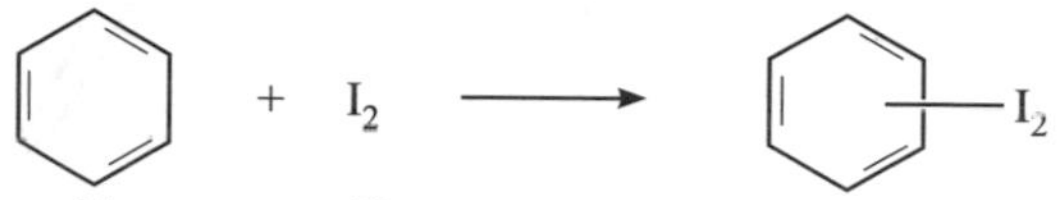

21 **22** **23**

Scheme 4

Another example, when aniline is dissolved in chloroform and tetracyanoethylene is added to it, a deep blue solution is obtained. This is due to the formation of charge transfer complex that absorbs in the visible region at 610 nm (**Scheme 5**).

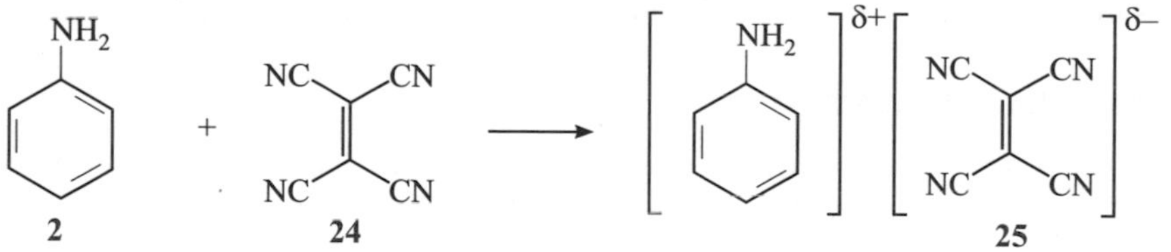

2 **24** **25**

Scheme 5

2.11.9 Study of Chemical Reactions

It has been used to follow the progress of a chemical reaction provided the absorption spectra of the reactant and product are considerably different. For example, the antioxidant activity using stable free radical can be analysed using this spectroscopic technique.

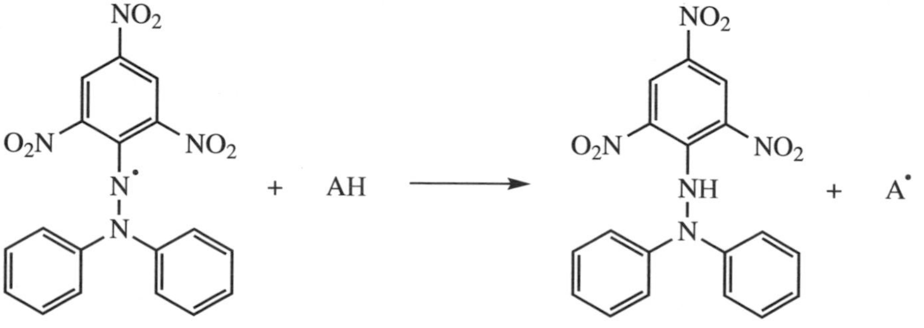

DPPH•, **26** DPPH-H, **27**

Scheme 6

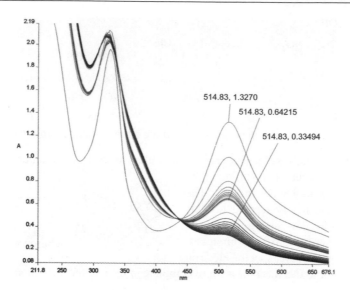

Figure 16. UV-Visible spectra indicating the reduction of DPPH˙ solution to DPPH-H in the presence of a hydrogen-donating antioxidant.

DPPH˙ (**26**, 1,1-diphenyl-2-picrylhydrazyl) has been used to evaluate the antioxidative activity of natural products in organic solvents. The molecule of DPPH˙ has deep violet colour, characterised by an absorption band in ethanol solution at 515–520 nm. When a solution of DPPH is mixed with a substance that can donate a hydrogen atom, then it gives rise to the reduced (**27**, DPPH-H) form with the loss of violet colour (**Scheme 6, Figure 16**).

2.11.10 Keto Enol Tautomerism

β-Diketone and β-keto ester are known to exist as tautomeric mixtures and their ultraviolet spectra will exhibit absorptions characteristic of both keto and enol forms, i.e., in addition to a weak n-π* absorption due to keto-carbonyl group, a strong π-π* due to conjugated double bond is also observed.

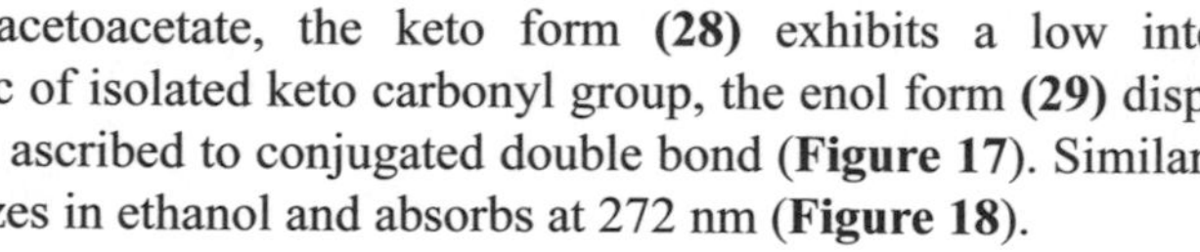

Keto-form, **28**   Enol-form, **29**

Figure 17. Tautomeric structures of ethylacetoacetate.

For example, ethylacetoacetate, the keto form (**28**) exhibits a low intensity band around 275 nm characteristic of isolated keto carbonyl group, the enol form (**29**) displays a high intensity band around 245 nm ascribed to conjugated double bond (**Figure 17**). Similarly, acetylacetone the keto form (**30**) enolizes in ethanol and absorbs at 272 nm (**Figure 18**).

Keto-form, **30**   Enol-form, **31**

Figure 18. Tautomeric structures of acetylacetone.

SOLVED PROBLEMS

Q1. What changes does absorption of UV-Visible radiation causes in the molecule?

Sol. Absorption of ultraviolet light molecule results in the electronic transitions, such as σ-σ*, π-π*, n-σ*, n-π*, leading to the promotion of electrons from lower energy level to higher energy level.

Q2. Predict the various electronic transitions possible in the following compounds.

    (i) CH_3CH_3 (ii) CH_3Br

    (iii) CH_3COCH_3 (iv) $CH_2=CH_2$

Sol. (i) σ-σ*   (ii) σ-σ* and n-σ*

    (iii) σ-σ*, n-σ*, π-π* and n-π*   (iv) σ-σ* and π-π*

Q3. Antibonding orbitals are described by the symbols σ* and π* and not by n*. Explain.

Sol. Since n-electrons are not engaged in bonding there are no corresponding antibonding (n*) orbitals.

Q4. Electronic absorption bands are generally broad as compared to infrared. Explain.

Sol. Electonic spectra arise from transition between electronic energy levels accomapanied by changes in both vibrational and rotational states. Since, the wavelength of absorption is a measure of the seperation of the energy levels of the orbitals concerned, a transition between these levels, when gaseous sample is irradiated gives rise to fine structures consisiting of a number of closely, spaced lines which very often merge (due to solvent-solute interactions) to give a broad absorption band, when spectral measurements are carried out in solution.

Q5. λ_{max} of benzenoid bands shifts towards longer wavelength as the number of linearly fused benzene rings increases. Explain.

Sol. As the number of linearly fused benzene rings increases the number of conjugated π-orbitals also increases, and the energy gap between HOMO and LUMO decreases resulting in the shifting of the bands towards longer wavelength.

Q6. Which of the following diene will absorb at a higher wavelength and why?

(a)

(b)

Sol. In both dienes there are four ring residues as substituents. But diene (b) has two exocyclic double bonds that contributes 10 nm and hence it absorb at higher wavelength.

(a) Acyclic butadiene   = 217 nm

    4 Ring residue (4 × 5)   = +20 nm

    Calculated value   = 237 nm

(b) Acyclic butadiene   = 217 nm

    4 Ring residue (4 × 5)   = +20 nm

    2 Exocyclic C=C (2 × 5) = +10 nm

    Calculated value   = 247 nm

Q7. Hydrogenation of following triene, 'A' with one equivalent of H_2 could give three isomers of $C_{10}H_{14}$. Show, how UV can distinguish these isomers.

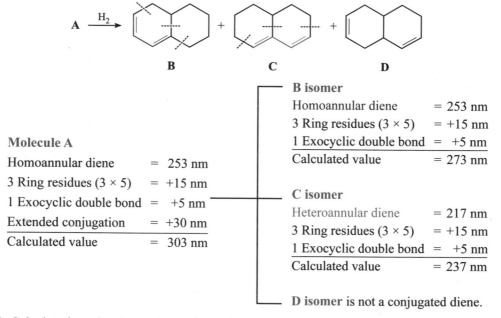

A

Sol. The molecule A undergoes reduction to give three isomers B, C, D. B and C are conjugated dienes they will absorb at 273 nm, 237 nm. D is not a conjugated diene, therefore, it will absorb at lower λ_{max} in comparison to B and C in the UV-Visible spectrum.

$$A \xrightarrow{H_2}$$

B   **C**   **D**

B isomer

Homoannular diene   = 253 nm

3 Ring residues (3 × 5)   = +15 nm

1 Exocyclic double bond = +5 nm

Calculated value   = 273 nm

Molecule A

Homoannular diene   = 253 nm

3 Ring residues (3 × 5)   = +15 nm

1 Exocyclic double bond = +5 nm

Extended conjugation   = +30 nm

Calculated value   = 303 nm

C isomer

Heteroannular diene   = 217 nm

3 Ring residues (3 × 5)   = +15 nm

1 Exocyclic double bond = +5 nm

Calculated value   = 237 nm

D isomer is not a conjugated diene.

Q8. Calculate λ_{max} for the π-π* transitions in the following compounds. Applying Woodward-Fieser rules. λ_{max} for heteronuclear diene is 215 nm, homoannular diene is 253 nm, one double bond exocyclic to 1 ring is 5 nm, two double bonds exocyclic to 2 rings is 10 nm. Alkyl ring residue is 5 nm.

(i) A B   (ii) C A B

Sol. Calculations of λ_{max} for compound (i) and (ii) is as follows:

(i)

Heteroannular diene	= 215 nm
1 Alkyl group	= +5 nm
3 Ring residues (3 × 5)	= +15 nm
1 Exocyclic double bond	= +5 nm
Calculated λ_{max}	= 240 nm

(ii)

Homooannular diene	= 253 nm
4 Ring residues (3 × 5)	= +20 nm
2 Exocyclic double bond (2 × 5)	= +10 nm
Calculated λ_{max}	= 283 nm

Q9. Acetone absorbs at 278 nm in hexane, whereas in water it absorbs at 265 nm. Explain.

Sol. Due to hydrogen bonding, which lowers the energy of the n-orbitals leading to hypsochromic shift.

UNSOLVED PROBLEMS

1. Explain the basic principle of UV spectroscopy.

2. Why UV-Visible spectroscopy is also called electronic spectroscopy?

3. What are the different regions and their ranges in UV-Visible spectroscopy?

4. What do you mean by vacuum UV-region? What is its range?

5. What type of compounds generally respond towards UV-Visible spectroscopy? Explain with example.

6. Define the following terms: (i) Absorbance, (ii) Transmittance, (iii) Absorption maxima (λ_{max}), (iv) Molar extinction coefficient (ε_{max}).

7. State and explain Beer-Lambert's law. What are its limitations?

8. What are the different electronic transitions possible in electronic spectroscopy?

9. What are the wavelength ranges for the ultraviolet and visible regions of the spectrum?

10. What molecular or structural features give rise to absorption of ultraviolet/visible (UV/Vis) radiation in organic species? Give an example of an organic compound that would not absorb UV/Vis radiation.

11. Why absorption bands are broad in UV-Visible spectroscopy?

12. Describe briefly how an ultraviolet spectrum can be scanned for pure organic compound?

13. What is electronic spectroscopy and its absorption range? Write the relationship between wavelength, frequency and wavenumber.

14. What do you mean by molar extinction coefficient (ε_{max})?

15. Define transmittance? What is its unit?

16. Why is the absorbance a function of C and l ?

17. Biphenyl exhibits a very intense band (ε_{max} = 19000) at 252 nm but its 2,2'-dimethyl derivative shows absorption similar to o-xylene (λ_{max} = 262 nm, ε_{max} = 270). Explain.

18. Calculate λ_{max} for the $\pi \rightarrow \pi^*$ transition for the following compounds using Woodward Fieser rules. λ_{max} for heteroannular diene is 215 nm, homoannular diene is 253 nm, one double bond exocyclic to 1 ring is 5 nm, two double bond exocyclic to 2 ring is 10 nm and alkyl ring residue is 5 nm.

(i)  (ii)

19. Calculate the λ_{max} value for the following (base value for s-cis diene = 253 nm, α,β-unsaturated carbonyl compounds = 215 nm)

(i) 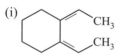 (ii)

20. Define bathochromic and hypsochromic shifts with appropriate examples.

21. What do you understand by the terms chromophore and auxochrome, describe with suitable examples.

22. How do you explain the fact that λ_{max} for compounds III and IV are vastly different, whereas those of I and II are very similar?

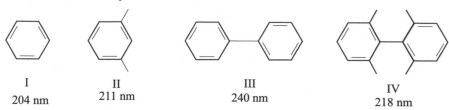

I	II	III	IV
204 nm	211 nm	240 nm	218 nm

23. How can you distinguish between the following three isomeric acids by UV spectroscopy? Use Woodward-Fieser rules to predict each λ_{max}.

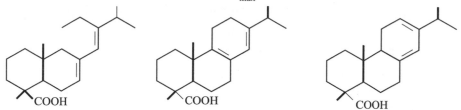

24. A diene $C_{11}H_{16}$ was thought to have the structure below. Its UV spectrum showed a λ_{max} of 263 nm. Can the structure below be correct? If not, draw a structure with the same skeleton that satisfies the spectral data.

25. Suggest the possible structures for the isomers with molecular formula C_4H_6O whose UV spectra show a high intensity peak at $\lambda_{max} = 187$ nm and very low intensity peak at $\lambda_{max} = 280$ nm.

26. Predict whether two isomers (a) and (b) can be distinguished using UV-Visible spectroscopy.

(a)   (b)

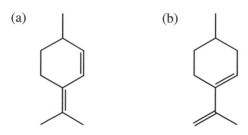

27. Dehydration of a tertiary alcohol (c), can give three possible conjugated dienes (e), (f) and (g). Give the structures of three dienes. Predict whether three products can be distinguished from each other on the basis of UV-Visible spectroscopy.

(c)   CH₃

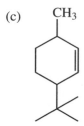

28. How the intensity of absorption of UV-band varies?

29. How steric hindrance affects the UV spectral absorption?

30. How will you distinguish between the K-bands of conjugated di and polyene system with those of enone system?

31. Saturated acyclic ketones usually record three absorption bands in their UV-spectra at around 160, 190 and 280 nm. Assign them in terms of electronic transitions. Predict with proper reasoning, which one will record the most intense absorption?

32. 1,3-Butadiene absorbs at longer wavelength, when it is irradiated with UV-radiation, in comparison to ethylene. Give reason.

33. The position of UV absorption maxima of aniline in aqueous solution are different from those of benzene but are almost identical with those in a solution of pH = 1. Explain.

34. Azobenzene is deep orange-red while hydrazobenzene is a colourless compound. Explain.

35. Does alkyl substitution or ring residue play any role on the absorption maxima of chromphore?

36. Acetyl acetone in aqueous solution records a UV absorption band at 274 nm (ε_{max} 2050), intensity of which differs markedly when recorded in isooctane 272 nm (ε_{max} 12,000). Justify this absorption.

37. In alcoholic solution the UV spectra of p-hydroxy acetophenone and p-methoxy acetophenone are virtually identical. What method will you follow so that the two compounds can be distinguished by UV-spectroscopy?

38. Explain, why the solution of iodine in hexane becomes brown when poured into benzene.

Chapter

3

Infrared Spectroscopy

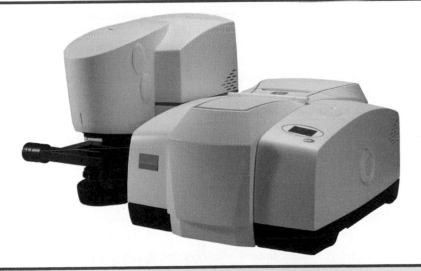

On 11 February 1800, Herschel was testing filters for the Sun so he could observe Sun spots. Herschel discovered infrared radiation in sunlight by passing it through a prism and holding a thermometer just beyond the red end of the visible spectrum.

In 1802, William Hyde Wollaston built a spectrometer.

Chapter 3

Infrared Spectroscopy

3.1 INTRODUCTION

Infrared spectroscopy (IR spectroscopy) deals with the infrared region of the electromagnetic spectrum. It is one of the several spectroscopic techniques applied for identification of molecular compounds. Which is based on the fact that different chemical functional groups absorb infrared light at different wavelengths depending upon the nature of the respective functional group. Except the enantiomers, no two compounds can have exactly the same IR spectrum, therefore the peak-by peak correlation of an unknown compound against an authentic sample can be used to identify the compound. Molecules are made up of atoms, which have a spherical symmetry and can have only electronic transitions. The moment two atoms combine to form a molecule, atoms loses their spherical symmetry to become elongated and allows the molecule to rotate and vibrate. These are associated with their quantised energy levels called electronic, vibrational and rotational energy levels. Each electronic level within a molecule is associated with a number of vibrational levels with less energy separations and each vibrational level in turn is associated with a set of rotational levels with even less energy separations. The transition between these levels gives rise to spectra with energy as E_{rot}, E_{vib} and E_{elect}. These energies have an order as:

$$E_{rot} < E_{vib} < E_{elect}$$

When a molecule absorbs radiation, its energy increases in proportion to the energy of the photon as per the relation:

$$E = h\nu = hc/\lambda$$

Where 'h' is the Planck's constant, ν is the frequency, λ is the wavelength and c is the velocity of light.

IR radiation posses relatively smaller amount of energy and therefore it induce transitions between vibrational and rotational energy levels of a molecule and hence it is also called as vibrational-rotational spectrum of the molecule.

3.2 HOOKE'S LAW

If in a diatomic molecule, atoms are held together by a mass less spring, it vibrates as follows the molecule from its equilibrium state gets stretched, come back to equilibrium position, gets compressed and comes back to equilibrium state that makes one vibration, the frequency of this vibration is given by:

$$\nu = \frac{1}{2\pi}\sqrt{\frac{K}{\mu}}$$

$$\bar{v} = \frac{1}{2\pi c} \sqrt{\frac{K}{\mu}}$$

v = frequency in Hz

$\bar{v}$ = wavenumber in cm^{-1}

K = force constant (dynes/cm)

μ = reduced mass in kg

c = velocity of light (3×10^{10} cm s^{-1})

Where, $\mu = \dfrac{m_1 m_2}{(m_1 + m_2)}$, masses of atoms in grams

or $\dfrac{M_1 M_2}{(M_1 + M_2)(6.02 \times 10^{23})}$, masses of atoms in amu

In polyatomic molecules, the molecular vibrations are more complex than diatomic molecules, as complex molecules have a large number of vibrational modes (including overtones, combinations and so called forbidden bands). However some vibrational modes can be attributed to functional groups and others to vibration of the whole molecular structure. Those vibrational modes (characteristic group vibration) can be best described mathematically if we imagine the two bonded atoms as two vibrating masses connected by a spring (a simple harmonic oscillator). The frequency of the vibration for this system is given by Hooke's law. Removing Avogadro's number (6.02×10^{23}) from denominator of the reduced mass expression (μ) by taking its square root, we obtain the expression:

$$\bar{v} = \frac{7.76 \times 10^{11}}{2\pi c} \sqrt{\frac{K}{\mu}} = \frac{7.76 \times 10^{11}}{2 \times 3.14 \times 3 \times 10^8} \sqrt{\frac{K}{\mu}}$$

$$= 4.12 \sqrt{\frac{K}{\mu}} \qquad \text{where } \bar{v} \text{ is in cm}^{-1}$$

$$\mu = \frac{M_1 M_2}{(M_1 + M_2)}$$

Where, M_1 and M_2 are atomic weights

K = force constant in dynes/cm [1 dyne = 1.020×10^{-3} g]

The above equation describes the major factors that influence the stretching frequency of a covalent bond between two atoms of mass m_1 and m_2 respectively. The force constant (K) is proportional to the strength of the covalent bond linking m_1 and m_2. For example, a C=N double bond is about twice as strong as a C–N single bond, and the C≡N triple bond is similarly stronger than the double bond. The IR stretching frequencies of these groups vary in the same order, ranging from 1100 cm^{-1} for C–N, to 1660 cm^{-1} for C=N, to 2220 cm^{-1} for C≡N. This equation may be used to calculate the appropriate position of a band in the IR spectrum by assuming that K for single, double, and triple bonds is 5, 10, and 15×10^5 dynes/cm, respectively. However, experimental and calculated values may vary considerably owing to resonance, hybridisation, and other effects which can operate in organic molecules.

Example 1: C=C bond

$$\bar{v} = 4.12 \sqrt{\frac{K}{\mu}} \qquad\qquad K = 10 \times 10^5 \text{ dynes/cm}$$

$$\mu = \frac{M_c M_c}{(M_c + M_c)} = \frac{(12)(12)}{12 + 12} = 6$$

$$\overline{v} = 4.12\sqrt{\frac{10 \times 10^5}{6}} = 1682 \text{ cm}^{-1} \text{ (calculated)}$$

$$= 1650 \text{ cm}^{-1} \text{ (experimental)}$$

Example 2: C–H bond

$$\overline{v} = 4.12\sqrt{\frac{K}{\mu}} \quad K = 5 \times 10^5 \text{ dynes/cm}$$

$$\mu = \frac{M_c M_H}{(M_c + M_H)} = \frac{(12)(1)}{12 + 1} = 0.923$$

$$\overline{v} = 4.12\sqrt{\frac{5 \times 10^5}{0.923}} = 3032 \text{ cm}^{-1} \text{ (calculated)}$$

$$= 3000 \text{ cm}^{-1} \text{ (experimental)}$$

3.3 INSTRUMENTATION

3.3.1 Fourier Transform Infrared Spectroscopy (FTIR)

Fourier Transform Infrared (FT-IR) spectrometry was developed in order to overcome the limitations encountered with dispersive instruments. Infrared energy is emitted from a glowing black-body source. This beam passes through an aperture which controls the amount of energy presented to the sample. The beam enters the interferometer where the "spectral encoding" takes place. The resulting interferogram signal then exits the interferometer. The beam enters the sample compartment where it is transmitted through or reflected off of the surface of the sample, depending on the type of analysis being accomplished. This is where specific frequencies of energy, which are uniquely characteristic of the sample, are absorbed. The beam finally passes to the detector for final measurement. The detectors used are specially designed to measure the special interferogram signal. The measured signal is digitised and sent to the computer where the Fourier transformation takes place. The final IR spectrum is then presented to the user for interpretation (**Scheme 1**).

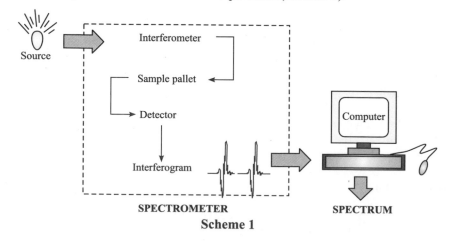

SPECTROMETER   **SPECTRUM**

Scheme 1

Advantages

Some of the major advantages of FT-IR over the dispersive technique is that it requires few seconds rather than several minutes to obtain spectra. It is highly sensitive that enable identification of even the smallest amount of contaminants and lowers the noise levels. These instruments employ a Helium-Neon laser as an internal wavelength calibration standard (referred to as the Connes Advantage). These instruments are self-calibrating and never need to be calibrated by the user.

3.4 SELECTION RULES

Each atom has three degrees of freedom, corresponding to motions along any of the three Cartesian coordinate axes (x, y, z). A polyatomic molecule of 'n' atoms has '3n' total degrees of freedom. However, 3 degrees of freedom are required to describe translational motion of the entire molecule through space. Additionally, 3 degrees of freedom correspond to rotation of the entire molecule. Therefore, the remaining 3n–6 degrees of freedom are fundamental vibrations for nonlinear molecules, whereas linear molecules possess 3n–5 fundamental vibrational modes, because only 2 degrees of freedom are sufficient to describe rotation. Infrared spectroscopy is based on the fact that molecules have specific frequencies at which they rotate or vibrate corresponding to discrete energy levels (vibrational modes). All molecular vibrations do not lead to observable infrared absorptions. Among the 3n–6 or 3n–5 fundamental vibrations, *those that produce a net change in the dipole moment may result in an IR activity and those that give polarisability changes may give rise to Raman activity.* Therefore, some vibrations can be both IR and Raman-active (**Figure 1**).

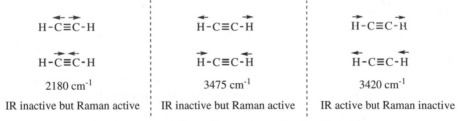

H–C≡C–H	H–C≡C–H	H–C≡C–H
H–C≡C–H	H–C≡C–H	H–C≡C–H
2180 cm^{-1}	3475 cm^{-1}	3420 cm^{-1}
IR inactive but Raman active	IR inactive but Raman active	IR active but Raman inactive

Figure 1. Conditions of a compound to be IR active.

The total number of observed absorption bands is generally different from the total number of fundamental vibrations. It is reduced because some modes are not IR active and a single frequency can cause more than one mode of motion to occur. Conversely, additional bands are generated by the appearance of overtones (integral multiples of the fundamental absorption frequencies), combinations of fundamental frequencies, differences of fundamental frequencies, coupling interactions of two fundamental absorption frequencies, and coupling interactions between fundamental vibrations and overtones or combination bands (Fermi resonance). The intensities of overtone, combination, and difference bands are less than those of the fundamental bands. The combination and blending of all the factors thus create a unique IR spectrum for each compound.

═══════ EXERCISE PROBLEMS 3.1 ═══════

1. What is the basic principle of infrared spectroscopy?
2. What is Hooke's law? What are the factors that influence the stretching frequency of a covalent bond between two atoms of mass m_1 and m_2.
3. What is Fourier transform infrared Spectroscopy? What are its advantages?
4. What are the necessary conditions for the compound to be IR active? Identify the diatomic molecules that do not absorb in the IR from the following: HCN, N_2, O_2, H_2, HCl.

3.5 SOME USEFUL TERMS IN INFRARED SPECTROSCOPY

3.5.1 Infrared Region of the Electromagnetic Spectrum

The infrared portion is divided into three regions; the near, mid and far-infrared **(Figure 2)**.

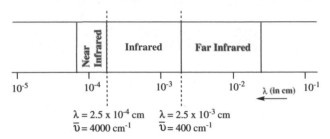

Figure 2. The electromagnetic spectrum representing different IR regions.

The far-infrared, approximately corresponds to 400–10 cm^{-1}, lying adjacent to the microwave region, has low energy and may be used for rotational spectroscopy. The mid-infrared, approximately 4000-400 cm^{-1} may be used to study the fundamental vibrations and associated rotational-vibrational structure. The higher energy near-infrared, approximately 14000-4000 cm^{-1} can excite overtone or harmonic vibrations.

3.5.2 IR Spectrum

IR spectrum of a compound represents its energy absorption pattern in the IR region and is obtained by plotting the percentage absorbance (transmittance) as a function of wavelength (wavenumber) over a particular range.

3.5.3 Relationship between Wavenumber and Wavelength

Wavenumber can be related to the frequency by using the following equation:

$$\text{Wavenumber} = \frac{1}{\text{Wavelength } (\lambda)}$$

The unit for wavenumber is cm^{-1}, therefore wavelength should be converted to cm rather than metre for this calculation. Wavelength is related to frequency as:

$$\text{Velocity of light (c)} = \frac{\text{Frequency (v)}}{\text{Wavelength } (\lambda)}$$

3.5.4 Relationship between Absorbance and Transmittance

Band intensities can be either expressed as transmittance (T) or absorbance (A). Absorbance is the ratio of the radiant flux absorbed by a body to that incident upon it.

$$A = \log(I_0/I)$$

Where, I_0 and I are the intensities of light before and after interaction with the sample respectively. Transmittance is the radiant power transmitted by a sample to the radiant power incident on the sample.

$$T = I/I_0$$

Therefore absorbance is the logarithm, to the base 10, of the transmittance:

$$A = \log_{10}(I/T)$$

3.5.5 Functional Group and Fingerprint Region

Absorption bands in the 4000 to 1500 cm⁻¹ region exhibits absorption bands corresponding to a number of functional groups and is called as the **'functional group region' (Scheme 2)**.

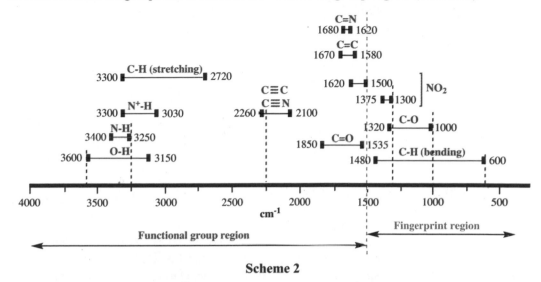

Scheme 2

The complexity of IR spectra in the region 1500 cm⁻¹ to 600 cm⁻¹ region makes it difficult to assign all the absorption bands, because it contains, apart from fundamental stretching vibrations, many bands resulting from the sum or difference of their vibrational frequencies. Thus, this part of the spectrum is the characteristic of a compound and because of the unique patterns found therein, it is often called the **'fingerprint region' (Scheme 2)**.

3.6 ABSORPTION OF INFRARED RADIATION

Infrared light is absorbed when oscillating dipole moment (due to molecular vibrations) interacts with the oscillating electric vector of the infrared beam. These vibrations absorb infrared light at certain quantised frequencies and give rise to characteristic bands. When the light of that frequency is incident on the molecule, energy is absorbed and the amplitude of that vibration is increased. Absorption of infrared radiation corresponds to the energy changes in the range from 2-10 kcal/mol. Radiation in this energy range corresponds to the stretching and bending vibrational frequencies of the bonds in most covalent molecules.

3.6.1 Molecular Vibrations

The covalent bonds of molecules are not rigid, but are more like stiff springs that can be stretched and bent. At ordinary temperatures these bonds vibrate in a variety of ways, and the vibrational energies of molecules may be assigned to quantum levels in the same manner as are their electronic states.

Figure 3. Modes of molecular vibrations involved in IR spectroscopy (a) Stretching; (b) Bending.

Transitions between vibrational energy states may be induced by absorption of IR radiation, having photons of the appropriate energy. When a compound absorbs the IR radiations, it is set into vibrations resulting in the excitation of bond stretching and bending (**Figure 3**).

3.6.1.1 Stretching vibrations

During stretching vibrations, the distance between two atoms increases or decreases but the atom remains in the same bond axis. It requires more energy to stretch (or compress) a bond and therefore such vibrations occur at higher frequency ($E = h\upsilon$).

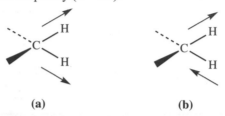

(a) (b)

Figure 4. Stretching vibration in IR spectrum (a) Symmetrical stretching; (b) Asymmetrical stretching.

There are two types of stretching vibrations:

(i) **Symmetrical stretching:** In this stretching mode, both the atoms move in and out simultaneously. For example, symmetrical stretching of R_2CH_2 group (**Figure 4**, **Table 1**).

(ii) **Asymmetrical stretching:** In this stretching mode, one atom moves in while the other moves out. For example, asymmetrical stretching of R_2CH_2 group (**Figure 4**, **Table 1**).

Table 1. Types of stretching vibration in different types of compounds.

Group	Symmetric stretching	Asymmetric stretching
Methyl	~2872	~2962
Anhydride	~1760	~1800
Amino	~3300 cm⁻¹	~3400 cm⁻¹

3.6.1.2 Bending vibrations (Deformations)

Bending vibrations can best be explained by considering the example of a three atom system including two terminal atoms bonded with a nodal one, wherein bonding distance between the atoms remains constant. But the position of the atoms effectively change relative to the original bond axis resulting in either decrease or an increase in the bond angle. It requires less energy to bend a bond and therefore such vibrations occur at lower frequency ($E = h\upsilon$). There are two types of bending vibrations (i) In plane bending vibrations (ii) Out of plane bending vibrations:

(i) In-plane bending vibrations: In the in-plane bending vibrations mode, the two terminal atoms tend to move towards or away from each other remaining in the same plane. For example, in-plane bending vibration (**Figure 5**). There are two types of in-plane bending vibrations: scissoring and rocking.

(a) Scissoring: In the scissoring vibrations both the atoms swing in concert toward opposite directions as represented for the group (**Figure 5a**). Imagine carbon as the pivot and the hydrogen's as the points of pair of scissors, then there is a plane passing through all the atoms and the motion is in plane.

(b) Rocking: In the rocking vibrations both the atoms swing to the same side and then both to the other side towards opposite direction with respect to the nodal atom as represented for the group (**Figure 5b**). Imagine the movement of the wipers of a car during rains.

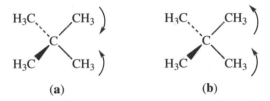

(a)   (b)

Figure 5. In plane bending vibrations in IR spectrum: (a) Scissoring; (b) Rocking.

(ii) Out-of-plane bending vibrations: In this bending vibration mode, both the atoms bend out of the nodal plane of the system, is called the out-of-plane bending vibration (**Figure 6**). Again there are two types of out-of-plane bending vibrations, i.e., wagging and twisting.

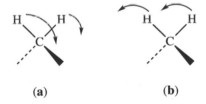

(a)   (b)

Figure 6. Out of plane bending vibrations in IR spectrum: (a) Wagging; (b) Twisting.

(a) Wagging: In the wagging vibrations both the atoms swing up and down out of the plane of paper (**Figure 6a**).

(b) Twisting: In the twisting vibrations one atom swings up and other swings down relative to the plane of paper (**Figure 6b**).

EXERCISE PROBLEMS 3.2

1. What is the infrared region of the electromagnetic spectrum?

2. What is the relationship between wavenumber and wavelength?

3. What is the relationship between absorbance and transmittance?

4. What is functional group and fingerprint region in infrared spectroscopy?

5. What is stretching vibrations?

6. Explain symmetric and asymmetric stretching?

7. What are bending vibrations?

3.7 FACTORS AFFECTING THE ABSORPTION FREQUENCIES

There can be many possible factors influencing the precise frequency of a molecular vibration, and it is usually impossible to isolate one effect from other.

3.7.1 Hybridisation

Hybridisation of atom-orbitals bears a great deal of significance in explaining the formation as well as the stability of molecular bonding imparting the related information such as bond length and bond strength. The types of hybridisation involving various atom-orbitals includes, sp, sp^2, sp^3 etc.

$$
\begin{array}{lll}
sp & sp^2 & sp^3 \\
\equiv\!C-H \ \ \text{Acetylinic} & =\!C-H \ \ \text{Vinyl} & C-H \ \ \text{Aliphatic} \\
& =\!C-H \ \ \text{Aromatic} & \\
& -\!C-H \ \ \text{Cyclopropyl} & \\
3300 \ \text{cm}^{-1} & 3100 \ \text{cm}^{-1} & 3000 \ \text{cm}^{-1}
\end{array}
$$

$\longleftarrow$

Stretching frequency increases with the increase in s-character

Figure 7. Effect of s-character on stretching frequency.

Greater is the s-contribution in the hybrid atomic orbitals of the valence state, the greater is the electronegativity of the atom. The electronegativity of the atom relative to a second atom is determined by the electronegativity of the hybridised atomic orbital with which they enter into bonding. Also, force constant K, is dependent on the hybridisation, higher is the s-character higher will be the strength of the bond and therefore higher will be the observed frequencies of C–H bond (**Figure 7**). The order of the bond strength is $sp > sp^2 > sp^3$.

3.7.2 Electronic Effects

3.7.2.1 Inductive effect

When an electronegative element is attached to an atom involved in the formation of a bond, it results in an increase in the absorption frequency of the atom. For example, highly electronegative halogen atom strengthens the carbonyl bond through an enhanced inductive effect and shifts the frequency to values even higher than that of esters (**Figure 8a**).

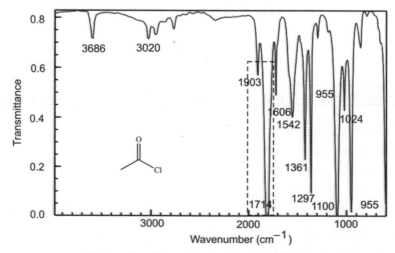

Figure 8a. IR spectrum of acetyl chloride.

However, when an electron donating group such as alkyl group is attached to carbonyl group the absorption of peak appears near 1665–1720 cm^{-1} (**Scheme 3, Figure 8b**).

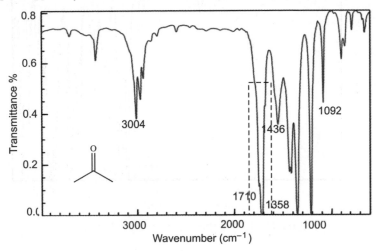

Figure 8b. IR spectrum of acetone.

Similarly, in anhydrides the frequencies are shifted to higher values than that of the esters because of the presence of electronegative oxygen atoms. Also, anhydrides give two absorption bands that are due to symmetric and asymmetric vibrations (**Figure 9, 10**).

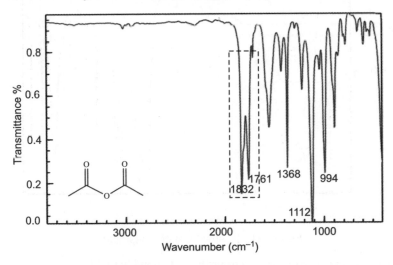

Figure 9. IR spectrum of acetic anhydride.

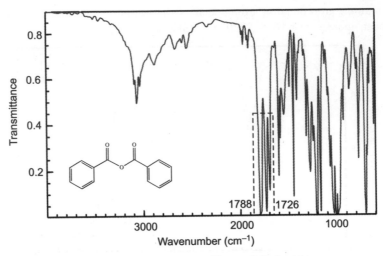

Figure 10. IR spectrum of benzoic anhydride.

3.7.2.2 Resonance effect

When a group containing the unpaired electron is in conjugation with the carbonyl group, it results in increased single bond character and a lowering of carbonyl frequency (**Figure 11**).

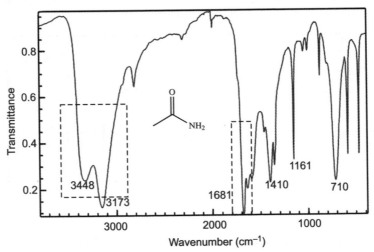

Figure 11. IR spectrum of acetamide.

In amides, due to the delocalisation of electrons the C–O bond has a high single bond character and C–N bond has a somewhat double bond character (**Scheme 3**).

Scheme 3

Similarly, if a carbonyl is conjugated with another double bond, the delocalisation of electrons of the carbonyl decreases its double bond character. Therefore, its absorption band shifts towards smaller wavenumber. For example, the absorption bands of α,β-unsaturated ketones (**Figure 12**) and those of the aromatic ketones (**Figure 13**) are situated at about 1682 and 1686 cm^{-1}, respectively, which is smaller than aliphatic ketones.

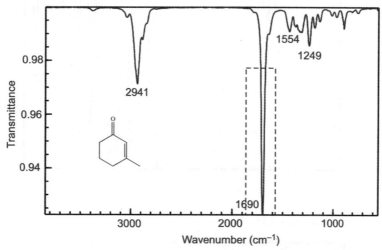

Figure 12. IR spectrum of methyl cyclohexenone.

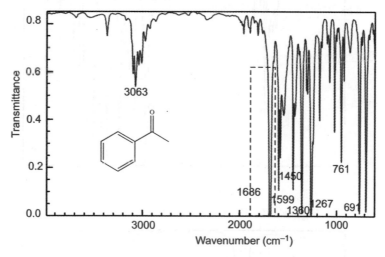

Figure 13. IR spectrum of acetophenone.

Similarly, the absorption bands of α,β-unsaturated esters and the aromatic esters (**Figure 14a** and **14b**), appears at a lower wavenumber than the aliphatic esters (**Figure 15**).

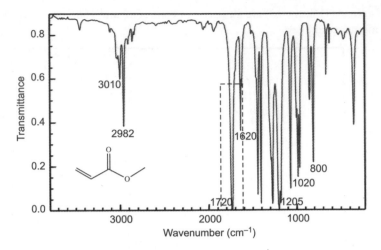

Figure 14a. IR spectrum of methyl acrylate.

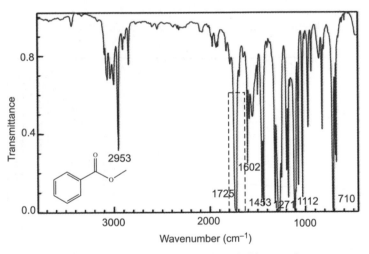

Figure 14b. IR spectrum of methyl benzoate.

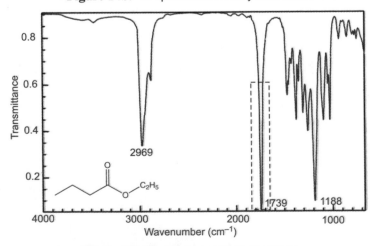

Figure 15. IR spectrum of ethyl butyrate.

3.7.3 Hydrogen Bonding

Hydrogen bonding especially in O–H and N–H highly influences the IR spectra. The use of polar solvents such as acetone has strong influence on the O–H and N–H absorptions. Alcohols, phenols and carboxylic acids in condensed phases are strongly H-bonded. They usually undergo intermolecular association leading to the formation of dimers, trimers and tetramers, which result in a wide range of absorptions and hence broadening of the absorption band. In dilute solution/ inert solvent (or in vapour phase) the proportion of free molecules increases that leads to absorption at a higher wavenumber. The H-bond can be regarded as a resonance hybrid of I and II, so that H-bonding involves a lengthening of the original O-H bond leading to the weakening of bond (its force constant is reduced), and lowering of the stretching frequency (**Scheme 4**).

Scheme 4

For example, carboxylic acid exists in monomeric form in vapour state, and absorbs at about 1760 cm^{-1}. However, acids in concentrated solution/neat liquid or in solid state tend to dimerise *via* hydrogen bonding. This dimerisation weakens (lengthens) the carbonyl bond and lowers the frequency to about 1710 cm^{-1}. Since there is a resonance form for the conjugated carbonyl group in which the C–O bond is a single bond. Thus, conjugated carbonyl group has a lower bond order than the isolated carbonyl group, so it has smaller K and hence a smaller stretching frequency (**Figure 16**).

3.7.4 Ring Strain and Size

Incorporation of the carbonyl group in a small ring (5, 4 or 3-membered) increases the stretching frequency. The increase in frequency ranges from 30 to 45 cm^{-1}, for a 5-membered ring, to 50 to 60 cm^{-1} for a 4-membered ring, and nearly 130 cm^{-1} for a 3-membered ring (**Figure 17**). For example, the absorption band of carbonyl group of strained cyclopropanone absorbs at 1815 cm^{-1}, while the unstrained cyclohexanone ring absorbs at about 1715 cm^{-1} (**Table 2**). In ketones larger rings have frequencies that ranges from nearly the same value as in cyclohexanone (1715 cm^{-1}) to values slightly less than 1715 cm^{-1}.

Table 2. Effect of ring strain on the C=O stretching in IR spectra.

Entry	Cyclic ketones	C=O stretching variation due to ring strain (cm^{-1})
1	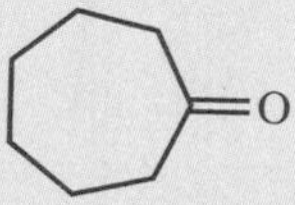	$1715 \rightarrow 1705$
2	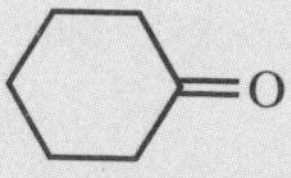	1715
3	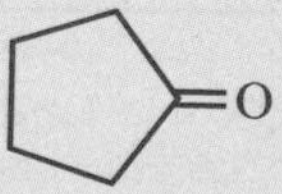	$1715 \rightarrow 1745$

Entry	Cyclic ketones	C=O stretching variation due to ring strain (cm^{-1})
4		1715 → 1780
5		1715 → 1815

3.7.5 Vibrational Coupling

Vibrational coupling takes place between two bonds vibrating with close frequency values, provided that the bonds are reasonably close in the molecule. The coupling vibration may be both fundamental (as in coupled stretching vibrations of AX_2 groups) or a fundamental vibration may couple with overtone of some other vibration. This later coupling is frequently known as Fermi resonance. For example, an isolated C–H has only one stretching frequency, but the stretching vibration of the $-CH_2$ group combine together to produce two coupled vibrations of different frequencies, anti-symmetric and symmetric combinations. The C–H bonds in CH_3 groups also give rise to such vibrations but these are of different frequencies than those of CH_2 groups.

3.7.6 Isotope Effect

The different isotopes in a particular species may give important information in IR spectroscopy. The mass effect on stretching frequencies is particularly evident when deuterium isotope equivalents are compared with corresponding hydrogen functions. Thus, the stretching frequency of a free O–H bond is 3600 cm^{-1}, but the O–D equivalent is lowered to 2600 cm^{-1}. If one of the bonded atoms (m_1 or m_2) is a hydrogen (atomic mass = 1), the mass ratio in the equation is roughly unity, but for two heavier atoms it is much smaller. Consequently, C–H, N–H and O–H bonds have much higher stretching frequencies than that of corresponding bonds to heavier atoms.

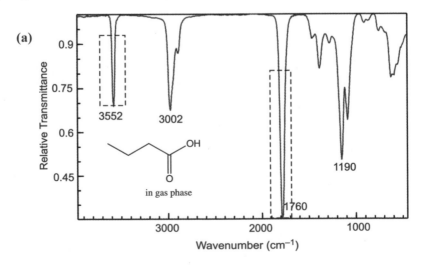

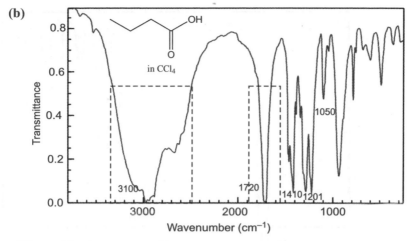

Figure 16. IR spectrum of butanoic acid, (a) In Gas phase; (b) In CCl₄.

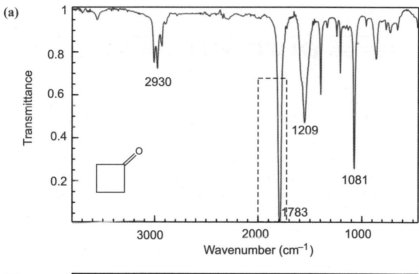

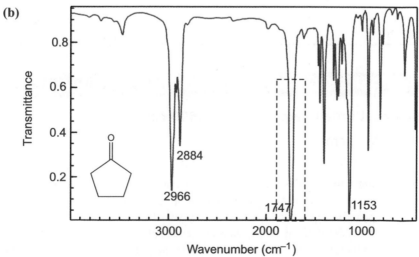

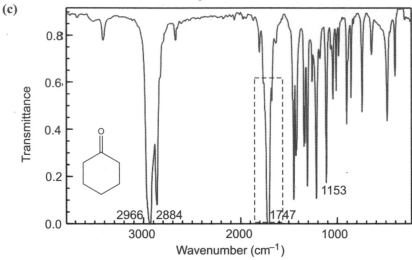

Figure 17. IR spectrum of cyclic ketones: (a) Cyclobutanone; (b) Cyclopentanone; (c) Cyclohexanone.

EXERCISE PROBLEMS 3.3

1. What are the factors affecting the absorption frequencies?
2. Explain the effect of hybridisation on the force constant of the bond.
3. How can you differentiate between 1-hexene and 1-hexyne on the basis of IR spectroscopy?
4. Explain the exceptionally low carbonyl stretching frequency in 2,4,6-cycloheptatrienone.
5. Why carboxylic acid exhibits a very broad band due to O–H stretching at 3000–2500 cm⁻¹ as compared to alcohols containing free OH group (3650–3600 cm⁻¹? Explain.
6. How hydrogen bonding affects IR frequency in an organic compound? How would you distinguish between intramolecular and intermolecular hydrogen bonding using IR spectroscopy?
7. How does the ring strain effect changes the position of absorption of C=O (stretching) in IR spectroscopy? Explain.
8. How conjugation changes the position of absorption of C=O (stretching) in IR spectroscopy? Explain.

3.8 CHARACTERISTIC GROUP VIBRATIONS

3.8.1 Functional Group Region (4000–1500 cm⁻¹)

3.8.1.1 4000–2300 cm⁻¹

The region represents vibrations of single bonds to H (O–H, N–H, N⁺–H, C–H, S–H).

(a) **O–H stretching (3600–2500 cm⁻¹):** O–H is present in phenols, alcohols and carboxylic acids, but the wavenumber is not reliable indicator of the functional group present, since the presence and position of O–H absorption is very sensitive to moisture and hydrogen bonding [**Figure 18(a) and 18(b)**]. Even the presence of a small amount of water in a sample gives a strong O–H absorption therefore the sample must be completely dry and moisture free. In general, the more concentrated solutions, the greater the extent of H-bonding and since H-bonding weakens the O–H bond, the absorption occurs at a lower frequency (**Table 3**).

(b) C–H stretching (3300–2750 cm^{-1}): C–H stretching vibrations are less useful, since all organic molecules contain C–H bonds and there is little difference in the stretching frequencies of the C–H bonds in most environments. The C–H stretching region, ranges from 3300–2750 cm^{-1}. *The frequency of the absorption of C–H bonds is a function mostly of the type of hybridisation of the bond.* The *sp* C–H bond present in the acetylinic compounds is stronger than the *sp^2* bond present in the vinyl compounds, which is stronger than the *sp^3* C–H absorption in saturated aliphatic compounds. This strength results in a larger vibrational force constant and a higher frequency of vibration. But the most useful C–H stretches are those for an aldehyde group and an alkyne. For example, in the IR spectrum of benzaldehyde there are two moderately strong C–H stretches at 2820 and 2738 cm^{-1} (**Table 3**).

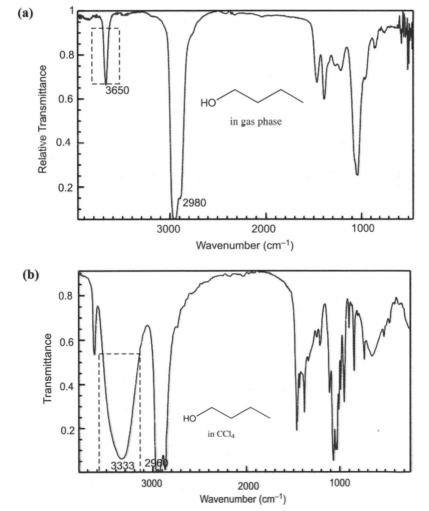

Figure 18. IR spectrum of *n*-butanol, (a) In gas phase; (b) In CCl$_4$.

3.8.1.2 N–H stretching (3500–3300 cm^{-1})

The N–H absorptions appear in the same region as that of O–H, but they are generally sharper because of the smaller degree of H-bonding. Primary amines [**Figure 19(a)** and **19(b)**] and primary amides can be easily distinguished from secondary amines and secondary amides [**Figure 20(b)**]. Primary amines and primary amides have two bands for symmetrical and asymmetrical stretches of the two N–H bonds, while secondary amines and secondary amides have only one N–H stretching vibration. However the tertiary amines do not have any N–H stretch (**Figure 21**, **Table 3**) N–H and O–H absorptions are observed nearly in the same region but for an N–H stretching vibration there should be a corresponding bending vibration around 1650–1550 cm^{-1}.

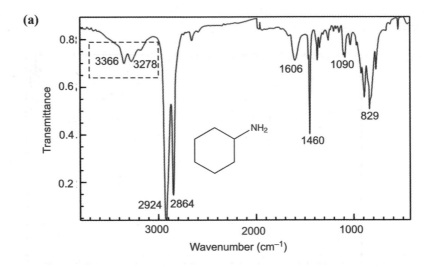

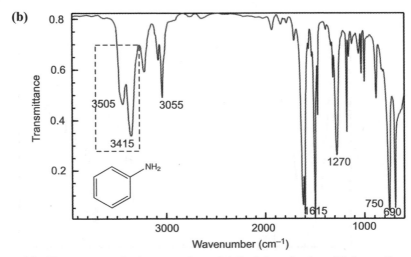

Figure 19. IR spectrum of primary amines, (a) Cyclohexylamine; (b) Aromatic amine.

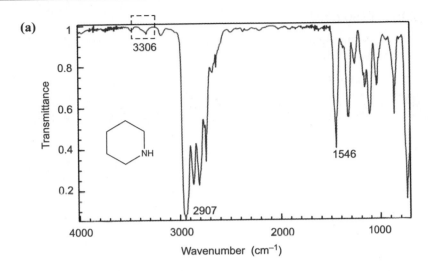

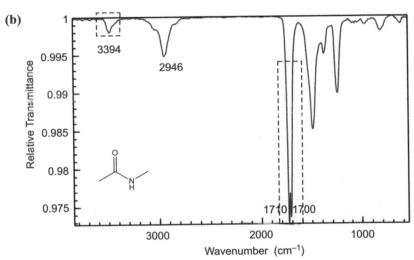

Figure 20. IR spectrum of (a) Piperidine; (b) N-methyl acetamide.

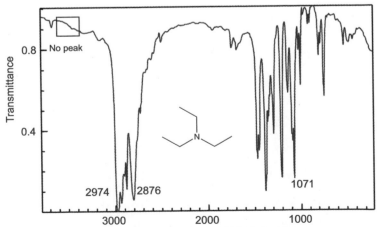

Figure 21. IR spectrum of triethylamine.

3.8.1.3   2300–1850 cm^{-1}

The region represents vibrations of triple and cumulative bonds of alkynes, nitriles, diazonium salts, azides, diazo compounds, carbodiimides, allenes, ketenes (**Table 4**).

Table 3. Single bonds to hydrogen values in IR spectrum.

Bonds	Wavenumber (cm^{-1})	Class of compound
C–H	3100–2700	Saturated alkanes
=C–H	3100–2700	Unsaturated alkenes or aromatic
≡C–H	3300	Terminal alkynes
O=C–H	2800 and 2700	Aldehydes (two weak peaks)
O–H (Free)	3600	Alcohols and phenols
O–H (H-bonded)	3400–3000	Alcohols and phenols (if H-bonding is present than a broad peak will be present at 3000–2500 cm^{-1}).
N–H	3450–3100	Amines (primary-several peaks, secondary-one peak, tertiary-no peak)
N$^+$–H	3300–2250	Quarternary ammonium salt (commonly shown by zwitterionic amino acids)
S–H	2700–2400	Thiol group

Table 4. The stretching vibrations for compounds containing multiple bond in IR spectrum.

Types of bond	Bonds	Wavenumber (cm^{-1})	Class of compound
Double bond	C=O	1840–1800 and 1780–1740	Anhydrides
		1815–1760	Acyl halides
		1750–1715	Esters
		1740–1680	Aldehydes
		1725–1665	Ketones
		1720–1670	Carboxylic acids
		1690–1630	Amides
	C=C	1675–1600	Often weak
	C=N	1690–1630	Often difficult to assign
	N=O	1560–1510 and 1370–1330	Nitro compounds
Triple bond	C≡C	2260–2120	Alkynes (bands are weak)
	C≡N	2260–2220	Nitriles

(a) –C≡C– stretching: The terminal C≡C of alkynes absorbs in the region 2140–2100 cm⁻¹, while the internal alkynes absorbs in the region 2260–2190 cm⁻¹ (**Figure 22a and 22b**). Most of these absorptions are moderate to weak. As the symmetry of the molecule increases the intensity of the alkyne band decreases.

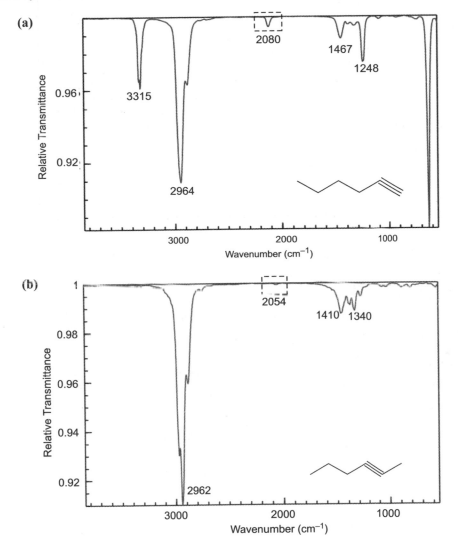

Figure 22. IR spectrum of (a) 1-Hexyne; (b) 2-Hexyne.

(b) –C≡N stretching: The absorption of nitrile appears in the region 2260–2100 cm⁻¹, which is stronger in nature (**Figure 23**).

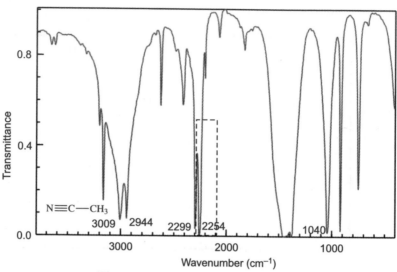

Figure 23. IR spectrum of acetonitrile

3.8.1.4   1850–1500 cm⁻¹

The region represents C=O, C=C and C=N vibrations, out of which the stretching vibrations for carbonyl groups are quite useful, as they are strong and very sensitive to the functional group attached to it.

(a) –C=O stretching (1850–1650 cm⁻¹): The carbonyl group is present in aldehydes, ketones, acid anhydrides, acid chlorides, esters, amides, imides, carbamates and carboxylic acids. This group absorbs strongly in the range from 1850–1650 cm⁻¹ because of its large change in dipole moment. Since the carbonyl stretching frequency is sensitive to attached atoms, the common functional groups absorb at characteristic values (**Scheme 5**).

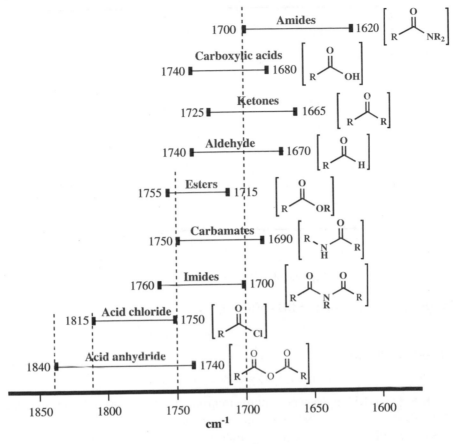

Scheme 5

The carbonyl frequency of ketone, which is approximately in the middle of the range, is usually considered as the reference point for the comparison of these values. The large range of the values may be explained through the use of inductive effect, resonance effect and hydrogen bonding. The first two effects operate in opposite way to influence the C–O stretching frequency.

(b) C=C and C=N stretching: These stretching vibrations are generally weaker than that of the carbonyls. The C=C stretching vibrations of aromatic ring occurs at around 1600 cm^{-1} **(Figure 24)**.

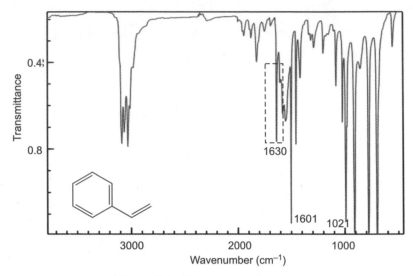

Figure 24. IR spectrum of styrene.

3.8.2 Fingerprint Region (1500–600 cm⁻¹)

This region mainly represents the vibrations of the whole skeleton of the molecule.

3.8.2.1 1500–1000 cm⁻¹

The region represents NO_2, N=O, C–O, S=O and P=O vibrations.

(a) **O–N=O, N=O stretching:** Generally, nitro group has two strong absorption bands in the region 1570–1500 cm⁻¹ and 1370–1300 cm⁻¹, while in case of nitroso (N=O) only one absorption band appears in the region 1600–1445 cm⁻¹. In nitrobenzene the asymmetric and symmetric stretching vibrations occur at 1525 and 1348 cm⁻¹ respectively.

(b) **C–O stretching:** The strong C–O absorption bands for alcohols, ethers, esters, carboxylic acids etc, appear in the region 1300–1050 cm⁻¹ (**Table 5**).

Table 5. Single bonds not attached to hydrogen and bending vibrations.

Single bonds	Bonds	Wavenumber (cm⁻¹)	Interpretation
Attached not to hydrogen	C–C	Variable	No diagnostic values
	C–O, C–N	1400–1000	Difficult to assign
	C–Cl	800–700	Difficult to interpret
	C–Br, C–I	Below 650	Often out of range of instrumentation
Bending vibrations	R–N–H	1650–1500	N–H bending should not be confused with C=O stretch in amides
	R–C–H	1480–1350	Saturated alkanes and alkyl groups
	R–C–H	1000–680	Unsaturated alkenes and aromatics

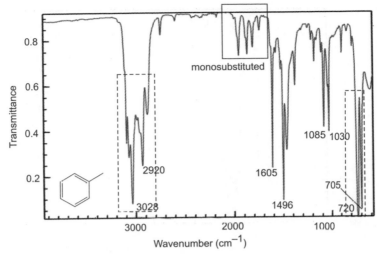

Figure 25. IR spectrum of toluene.

(a)

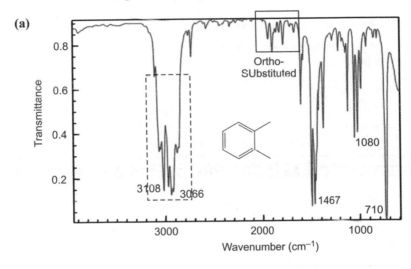

(b)

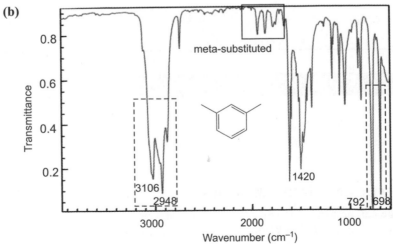

(c)

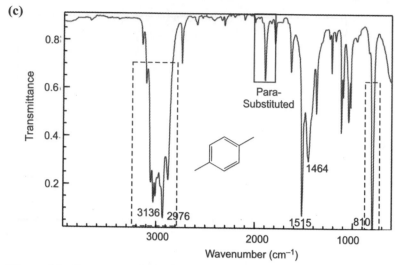

Figure 26. IR spectrum of (a) *o*-Xylene; (b) *m*-Xylene; and (c) *p*-Xylene.

3.8.2.2 1000–666 cm⁻¹ (Bending vibrations)

The vibrations present in the region can be attributed to out of plane bending vibrations of unsaturated C–H bending vibrations. The monosubstituted aromatic rings can be distinguished from disubstituted aromatic rings. The absorption of monosubstituted aromatic rings occurs in the region 770–730 cm⁻¹ and 710–690 cm⁻¹ **(Figure 25)**, whereas the 1,2-disubstituted aromatic rings show absorptions at 770 cm⁻¹–735 cm⁻¹ **[Figure 26(a)]**. Similarly, 1,3 and 1,4-disubstituted aromatic rings shows the absorptions in the region 810–750 cm⁻¹, 725–680 cm⁻¹ for former and 860–800 cm⁻¹ for later respectively **[Figure 26(b) and 26(c)]**.

═══════ EXERCISE PROBLEMS 3.4 ═══════

Give the approximate position of the characteristic IR bands in the following compounds:

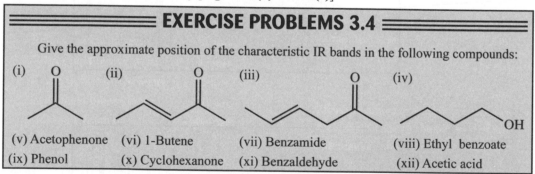

(v) Acetophenone   (vi) 1-Butene   (vii) Benzamide   (viii) Ethyl benzoate
(ix) Phenol   (x) Cyclohexanone   (xi) Benzaldehyde   (xii) Acetic acid

3.9 APPLICATIONS OF IR SPECTROSCOPY

This technique works almost exclusively on samples with covalent bonds. Simple spectra are obtained from samples with few IR active bonds and high levels of purity. More complex molecular structures lead to more absorption bands and more complex spectra. The technique has been used for the characterisation of very complex mixtures.

3.9.1 Detection of Functional Groups

This technique can be used to identify the presence of functional groups that absorb in the region 3500–1500 cm⁻¹. The absence of a band at a particular wavenumber may be regarded as an evidence for the absence of a particular group in the compound. For example, the oxygen in a compound can be present only as O–H, C=O, or C–O–C the presence or absence of the absorption in the carbonyl region (1870–1650 cm^{-1}) or hydroxyl region (3700–3200 cm^{-1}) can serve to ascertain some of these possibilities. It is thus possible to establish a correlation between infrared absorption and functional groups.

3.9.2 Effect of Different Types of Hydrogen Bonding

Intramolecular hydrogen bonding is independent of the concentration of the sample and so the position of the carbonyl band will not vary with concentration. For example, due to this effect a decrease in the carbonyl frequency in methyl salicylate can be observed (**Figure 27**).

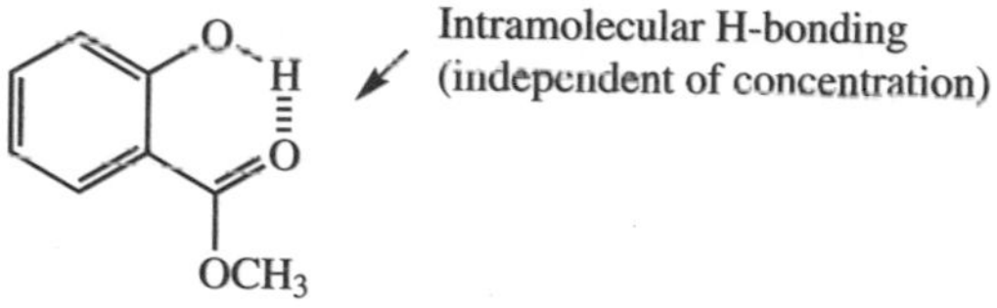

Intramolecular H-bonding
(independent of concentration)

Methyl salicylate (C=O stretching at 1680 cm^{-1})

Figure 27. Effect of intramolecular hydrogen bonding on vibrational stretching in methyl salicylate.

On the other hand intermolecular hydrogen bonding is dependent upon the concentration. The more concentrated the sample, the lower the stretching frequency (**Figure 28**) due to the greater extent of hydrogen bonding.

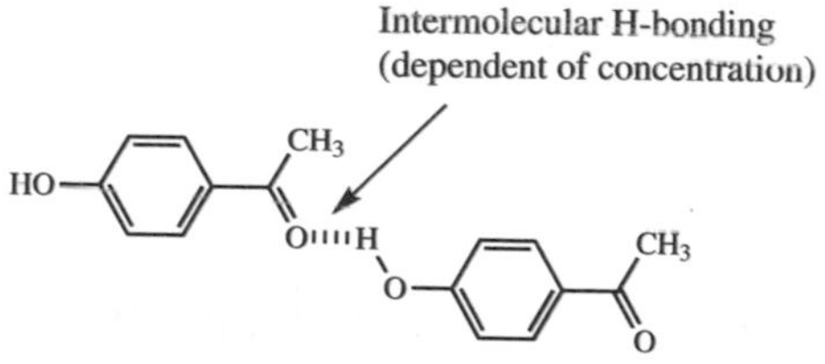

Intermolecular H-bonding
(dependent of concentration)

4'-hydroxyacetophenone

(i) 10 mg/cc solution in chloroform   (ii) 20 mg/cc solution in chloroform
(C=O stretching at 1672 cm^{-1})   (C=O stretching at 1665 cm^{-1})

Figure 28. Effect of intermolecular hydrogen bonding on vibrational stretching in 4'-hydroxyacetophenone.

3.9.3 Effect of Position of Substituent on Stretching Frequency

Axial and equatorial substituents on cyclohexane ring show different stretching frequencies.

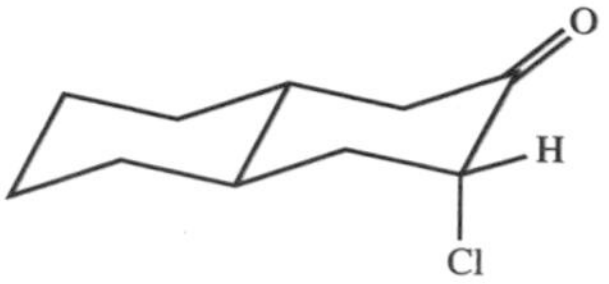

Axial chlorine atom
C=O stretching at ~1725 cm^{-1}

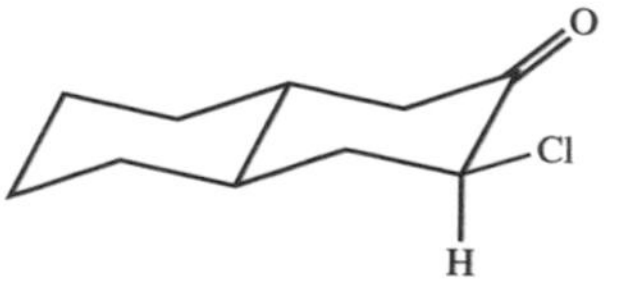

Equatorial chlorine atom
C=O stretching at ~1750 cm^{-1}

Figure 29. Effect of position of substituent on stretching frequency in the IR spectrum of cycloalkanes.

An equatorial substituent usually absorbs at a higher frequency than does the same substituent at axial position. The higher frequency of the C–X bond in equatorial position is due to less steric interaction of C–X bond with adjacent hydrogen atoms. For example, in case of the 2-chloro-1-decalone, when the chlorine atom is at axial position the carbonyl frequency is 1725 cm^{-1}, while at equatorial position it is 1750 cm^{-1} (**Figure 29**).

3.9.4 Study of Chemical Reactions

Infrared spectroscopy has been used to predict the product formed particularly in a photochemical reaction. Suitable constituted unsaturated ketones are liable to photocatalysed rearrangements. It has been anticipated that irradiation of naturally occurring verbenone might give known isomer, chrysanthenone. When verbenone is irradiated in ethanol solution, the UV absorption of verbenone gradually disappears and the IR spectrum of the crude products exhibit bands at, (a) 1685 cm^{-1}, (b) 1787 cm^{-1}, (c) 1740 cm^{-1} and (d) 1715 cm^{-1} (**Scheme 6**).

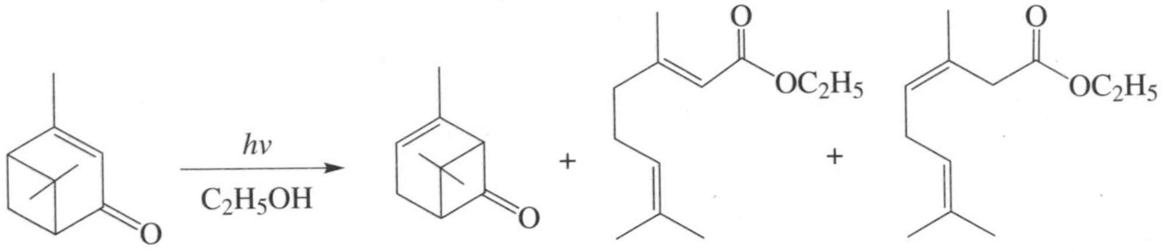

Verbenone (1685 cm^{-1})	Chrysanthenone (1787 cm^{-1})	Ethyl geranate (1715 cm^{-1})	Ethyl-3,7-dimethylocta-3,6-dienoate (1740 cm^{-1})

Scheme 6

3.9.5 Keto-Enol Tautomerism in Organic Compounds

β-diketones and β-keto esters are known to exist as tautomeric mixtures and their IR spectra will exhibit absorptions characteristic of both *keto* and *enol* forms, i.e., in addition to the carbonyl stretching frequency, a broad O–H stretching frequency and a C=C stretching frequency are also observed (**Figure 30**).

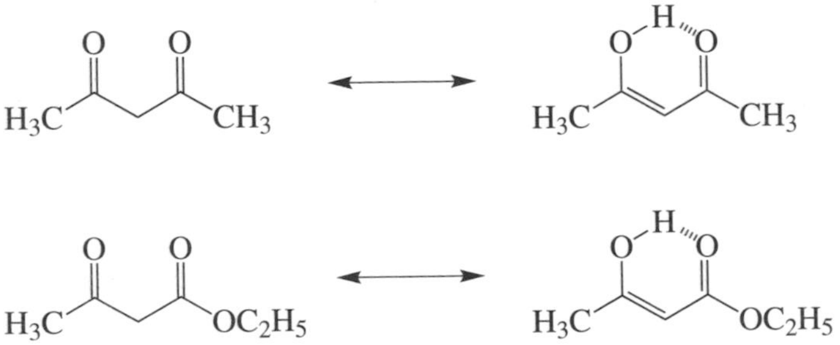

Figure 30. Keto-enol tautomerism in β-diketones and β-keto esters.

3.9.6 Geometrical Isomerism

The C=C stretching frequency in *trans*-1,2-dichloroethylene (1580 cm^{-1}) is not observed in its IR spectrum because the symmetric C–Cl stretching vibration produces no change in the dipole moment of the molecule and consequently there will be no absorption for C=C stretching vibration (**Figure 31**).

Cis μ = 1.85   Trans μ = 0

Figure 31. Effect of change in dipole moment on geometrical isomers of 1,2-dichloroethylene.

Similar, considerations apply to C≡C stretching mode in methylacetylene, the vibrations are infrared active and a strong band is observed at 2150 cm^{-1}, whereas in dimethylacetylene the vibrations are symmetrical about C≡C and consequently the band at 2150 cm^{-1} remains forbidden.

3.10 INTERPRETATION OF THE INFRARED SPECTRA

Identification of characteristic absorption bands caused by different functional groups is the basis of infrared spectra. **Table 6** can be utilised for the interpretation of infrared spectra.

Table 6. The absorption values of different functional groups in IR spectroscopy.

Class of compound	Structure	Type	Characteristic absorption (cm^{-1}) and type of vibrations
Alkane	R–CH$_2$–R	–C–H	3000–2850 (Stretching)
		–CH$_2$ and CH$_3$	1480–1440 (Bending) 1380–1370
Alkene	RHC=CH$_2$	=C–H	3140–3080 (Stretching)
		–C=C–	1680–1620 (Stretching)
	R$_2$C=CH$_2$	=C–H	3140–3080 (Stretching)
		C=C	1650–1645 (Stretching)
	cis RCH=CHR	=C–H	3020–3010 (Stretching)
		C=C	1660–1655 (Stretching)
		=C–H	1000–675 (Out-of-plane bending)
	trans RCH=CHR	=C–H	3020–3010 (Stretching)
		C=C	1675–1670 (Stretching)
		=C–H	980–970 (Out-of-plane bending)
	R$_2$C=CHR (trisubstituted)	C=C	1670–1660 (Stretching)
	R$_2$C=CR$_2$ (tetrasubstituted)	C=C	1670–1660 (Stretching)
	R$_2$C=CH–CH=CH$_2$	C=C	1660–1580 (Stretching)
Alkyne	RC≡CH	–C–H	3310–3300 (Stretching)
		–C≡C–	2140–2100 (Stretching)

Class of compound	Structure	Type	Characteristic absorption (cm^{-1}) and type of vibrations
		–C–H	700–600 (Bending)
	RC≡CR	–C≡C –	2260–2190 (Stretching)
Mononuclear aromatic hydro carbon	Benzene and its derivatives	Ar–H C=C C–H	3050–3000 (Stretching) 1600–1440 (Stretching) 900 – 700 (Bending)
	Monosubstituted	C–H	770–730 (Bending) 710–690 (Bending)
	o-Disubstituted	C–H	735–700 (Bending)
	m-Disubstituted	C–H	810–750 (Bending) 725–680 (Bending)
	p-Disubstituted	C–H	800–860 (Bending)
Carboxylic acid	Aliphatic carboxylic acid	C=O	1725–1700 (Stretching)
		C–O	1320–1210 (Stretching)
		O–H	3550–3500 (Stretching)
	RCH=CHCOOH (Unsaturated)	C=O	1715–1690 (Stretching)
	ArCOOH (Aromatic)	C=O	1700–1680 (Stretching)
		C=O (H-bonded)	1680–1650 (Stretching)
	XCH$_2$COOH (α-Halogeno acids)	C=O	1740–1715 (Stretching)
Aldehyde	R–CH=O (Aliphatic saturated)	C=O	1740–1720 (Stretching)
		C–H	2850–2820 (Stretching) 2750–2720 (Stretching) 975–780 (Bending)
	R–CH=CH–CH=O α,β-unsaturated aldehyde)	C=O	1705–1680 (Stretching)
	ArCHO	C=O	1715–1695 (Stretching)
Amide	–CONH$_2$ (Primary)	C=O	1690–1640 (Stretching)
		N–H	3500–3100 (Stretching)
		N–H	1640–1550 (Bending)
		C=O	1700–1670 (Stretching)
	–CONH (Secondary)	N–H	3340–3150 (Stretching)
		N–H	1560–1520 (Bending)
	–CONR$_2$ (Tertiary)	C=O	1660–1620 (Stretching)

Class of compound	Structure	Type	Characteristic absorption (cm^{-1}) and type of vibrations
Imides	Acyclic	C=O	1710 (Stretching)
	Cyclic (5-membered)	C=O	1760–1700 (Stretching)
	Cyclic (6-membered)	C=O	1710–1700 (Stretching)
Lactams	β-Lactam	C=O	1750–1720 (Stretching)
	γ-Lactam	C=O	1700 (Stretching)
	δ-Lactam	C=O	1680 (Stretching)
Anhydride	Acyclic	C=O	1830–1800 (Stretching) 1775–1740 (Stretching)
	Aromatic	C=O	1770–1720 (Stretching)
	Cyclic saturated (5-membered)	C=O	1870–1820 (Stretching) 1800–1750 (Stretching)
		C–O	1310–1210 (Stretching)
	α,β-Unsaturated	C=O	1850–1800 (Stretching) 1830–1780 (Stretching)
Ester	Saturated acyclic	C=O	1750–1735 (Stretching)
		C–O	1300–1000 (Stretching)
	–C=C–COOR	C=O	1730–1715 (Stretching)
	–CO–COOR	C=O	1755–1740 (Stretching)
	β-Keto ester	C=O	1655–1635 (Stretching)
	γ-Keto ester	C=O	1750–1735 (Stretching)
Ketone	Saturated acyclic	C=O	1725–1705 (Stretching)
	Cyclic	C=O	3-membered, 1850 (Stretching) 4-membered, 1780 (Stretching) 5-membered, 1745 (Stretching) 6-membered, 1715 (Stretching) 7-membered, 1705 (Stretching)
	α,β-Unsaturated ketone	C=O	1685–1665 (Stretching)
	Aryl ketone	C=O	1700–1680 (Stretching)
	α-Diketone	C=O	1730–1710 (Stretching)
	β-Diketone	C=O	1640–1535 (Stretching)
	γ-Diketone (–COCH$_2$CH$_2$CO–)	C=O	1725–1705 (Stretching)
	Ketone (R–C=C=O)	C=O	2150 (Stretching)
Alcohol	ROH (Primary)	O–H (H-bonded)	3600–3200 (Stretching)
		O–H	1250–1350 (Bending)
		O–H (free)	3700–3500 (Stretching)
		C–O	1150–1050 (Stretching)

Class of compound	Structure	Type	Characteristic absorption (cm⁻¹) and type of vibrations
	R₂CHOH (Secondary)	O–H (H-bonded)	3600–3200 (Stretching)
		O–H	1350–1250 (Bending)
		C–O	1120–1030 (Stretching)
	R₃COH (Tertiary)	O–H (H-bonded)	3600–3200 (Stretching)
		O–H	1410–1300 (Bending)
		C–O	1170–1100 (Stretching)
Phenol		O–H (H-bonded, Intermolecular)	3500–2400 (Stretching)
		O–H (H-bonded, Intermolecular)	3400–3200 (H-bonded Intramolecular)
		C–Br	600–500 (Stretching)
		C–I	500 (Stretching)
Lactones	Saturated-γ-lactone	C=O	1780–1700 (Stretching)
	Saturated-δ-lactone	C=O	1750–1735 (Stretching)
	α,β-Unsaturated-γ-lactone	C=O	1760–1740 (Stretching)
	β,γ-Unsaturated-δ-lactone	C=O	1805–1785 (Stretching)
Amine	Primary	N–H	3500–3000 (Stretching)
		C–N	1360–1080 (Stretching)
		N–H	1600 (Bending)
	Secondary	N–H	3500–3200 (Stretching)
		C–N	1360–1310 (Stretching)
		N–H	1580–1490 (Bending)
	Tertiary	C–N	1360–1310 (Stretching)
Imine	=NH	N–H	3400–3000 (Stretching)
Nitro	Aliphatic	N=O	1560–1540 (Stretching)
	Aromatic	N=O	1370–1360 (Stretching)
Azo	–N=N–	N=N	1645 (Stretching)
Nitrile	R–CN (Aliphatic, Saturated)	C≡N	2245 (Stretching)

Class of compound	Structure	Type	Characteristic absorption (cm^{-1}) and type of vibrations
	–C=C–CN Aliphatic, α,β-Unsaturated	$C\equiv N$	2220 (Stretching)
	ArCN Aromatic	$C\equiv N$	2245–2225 (Stretching)
Isocyanide	C–NC	$C\equiv N$	2220–2250 (Stretching)
Isocyanate	C–NCO	$C=N$	2270–2242 (Stretching)
Nitroso	C–NO	$N=O$	1600–1500 (Stretching)
Alkyl halide	R–X	C–F	1400–1000 (Stretching)
		C–Cl	800–600 (Stretching)
		C–Br	600–500 (Stretching)
		C–I	500 (Stretching)
Thiol	$R_2C=S$	C=S	1200–1050 (Stretching)

SOLVED PROBLEMS

Q1. The stretching of C–H bond in an alkane leads to absorption at 2900 cm^{-1} in IR spectrum. What wavelength does this corresponds to?

Sol. Since $\lambda = 1/v = 1/2900$ cm^{-1} = 3.45 × 10^{-4} cm^{-1}

Q2. Compare the relative stretching frequencies for C–C, C=C and C≡C.

Sol. The frequency of vibration is directly proportional to the force constant of the bond. Thus the increasing stretching frequency follows the order C–C (1200 cm^{-1}) < C=C (1650 cm^{-1}) < C≡C (2150 cm^{-1}).

Q3. Explain the exceptionally low carbonyl frequency in 2,4,6-cycloheptatrienone.

Sol. The exceptionally low carbonyl frequency in 2,4,6-cycloheptatrienone is due to conjugation effect. Because the cycloheptatrienyl cation, contains six π-electrons and consequently attains a stable aromatic structure, there are strong resonance contributions from the hybrid with a single-bond to oxygen resulting in an extremely low carbonyl frequency.

Q4. Arrange, giving reasons, the following lactones in order of their decreasing carbonyl frequency.

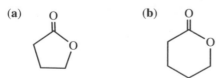

(a)   **(b)**

Sol. The ring size have a dramatic effect on the stretching frequency, therefore the five membered lactone (a), will absorb at higher frequency than six membered ring compound (b).

Q5. The carbonyl absorption of the carboxylate anion differs considerably from that of the parent acid. Explain.

Sol. The carboxylate group is symmetrical due to resonance and the two CO bands are equivalent strength, intermediate between C=O and C–O. The two bands, one near 1600 cm^{-1} and other near 1400 cm⁻¹, observed in carboxylate anion are due to symmetric and asymmetric stretching.

Q6. Which of the following molecule has the highest carbonyl stretching frequency.

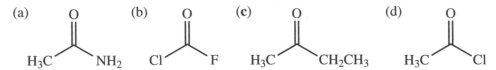

Sol. Except acetamide, the inductive effect is dominant and the highest carbonyl frequency will be shown by (b) because it involves the largest net inductive withdrawal from carbonyl group.

Q7. Deduce which of the following compounds would give a spectrum showing strong absorption at 1720 cm⁻¹.

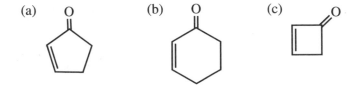

Sol. Since the C=O stretching frequency increases as the size of the ring decreases by introduction of olefinic bond in α,β-position, the ketone (b) cycloheptatrienone will show strong absorption at 1720 cm⁻¹.

Q8. How would you distinguish between the N–H stretching absorption of a primary amine and secondary amine.

Sol. N–H stretching absorption of a primary amine would appear as a doublet, the two components corresponding to asymmetric and symmetric modes. The separation is usually 100 cm⁻¹. N–H stretching absorption of a secondary amine would generally have a single maximum.

Q9. Amongst α,β-unsaturated ketone and aliphatic ketone, which will absorb at higher frequency and why?

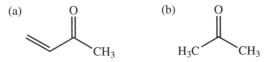

Sol. The α,β-unsaturated ketones absorbs at about 1682 cm⁻¹, while aliphatic ketones absorb nearly at 1700 cm⁻¹ which is due to the conjugation present in α,β-unsaturated ketones.

Q10. Carboxylic acid in vapour state absorbs at about 1760 cm⁻¹, whereas in concentrated solution/ neat liquid or in solid state it absorbs nearly 1680 cm⁻¹. Explain.

Sol. Carboxylic acid exists in monomeric form in vapour state, and absorbs at about 1760 cm⁻¹. However, acids in concentrated solution/neat liquid or in solid state tend to dimerise *via* hydrogen bonding leading to the reduction in absorption frequency.

UNSOLVED PROBLEMS

1. What is the principle of IR spectroscopy?

2. Write notes on: (i) Rocking, (ii) Wagging, and (iii) Scissoring vibrations?

3. How many fundamental vibrations do you expect for a molecule of C_6D_6?

4. How will you distinguish between acid chlorides, acid anhydrides, esters and amides on the basis of IR spectroscopy?

5. What do you mean by finger print and functional group region?

6. Explain, why?

   (a) O–H absorbs at a higher wavenumber than N–H.

   (b) Alkynes absorb at higher wavenumber than alkenes.

   (c) The carbonyl absorption of the carboxylate ion differs considerably from its parent acid.

   (d) C–H stretching vibrations move to higher frequency in the sequence:

   Alkane < Alkene < Alkyne

7. How IR spectroscopy be used to distinguish between inter and intra molecular hydrogen bonding?

8. What is the frequency of vibration in Hz (sec^{-1}) of the O–H bond? What is the wavelength of the light in nm that is absorbed by this bond? What is the energy, in Joules, of one photon of this light?

9. In each of the following pairs of bonds, select the one that stretches at the higher frequency.

   (a) C–O or C=O   (b) C–O or C–Cl   (c) C=C or C≡C

   (d) C–C or C–O   (e) C–H or O–H

10. Predict the approximate positions of all of the important absorptions in the IR spectrum of this compound.

$$CH_3CH_2CH=CHCOOH$$

11. Explain, how IR spectroscopy could be used to distinguish between the following two compounds.

12. An organic compound **A** shows the peak at 1679 cm^{-1} in the IR spectrum. Which of the following four compounds is the most suitable structure for this compound.

13. Explain, why IR spectrum of liquid t-butyl alcohol shows a strong absorption at 3360 cm^{-1}? Whereas a very dilute solution of the same compound in CCl_4 shows a strong absorption band at 3620 cm^{-1} instead of 3360 cm^{-1}.

14. How the following pairs of compounds can be distinguished using IR spectroscopy?

   (a) CH_3COCH_3 and CH_3CHO

   (b) CH_3COOH and CH_3OH

(c) $(CH_3)_2C=C(CH_3)_2$ and $CH_3CH_2CH=CHCH_3$

(d) CH_3CH_2COOH and CH_3COOCH_3

(e) $p\text{-}ClC_6H_4COCH_3$ and $C_6H_5CH_2COCl$

(f) $CH_3CH_2C\equiv CH$ and $CH_3CH_2C\equiv N$

(g) $C_6H_5NHCOCH_3$ and $p\text{-}CH_3C_6H_4CONH_2$

(h) Cyclohexanone and 3-methylcyclopentanone

(i) CH_3CH_2COOH and $CH_3CH_2COO^-$

(j) $C_6H_5NH_2$ and $C_6H_5CONH_2$

(k) CH_3COCH_3 and CH_3CH_2OH

(l) CH_3COOH and $CH_3CH_2COCH_3$

(m) CH_3COOH and CH_3CHO

(n) *cis*-Stilbine and *trans*-stilbine

(o) Cyclohexanol and cyclohexanone

15. Give the positions of the following characteristic absorptions in IR spectroscopy:

(a) C–H stretching in terminal alkyne

(b) C≡C stretching in alkyne

(c) How hydrogen bonding affect the vibration frequency of O–H group.

16. Why water and ethanol are not commonly used in IR spectroscopy?

17. 1-Acetyl-2-methylcyclohexene exhibits carbonyl stretching frequency slightly higher (1693 cm^{-1}) than that of 1-acetyl-cyclohexene (1686 cm^{-1}). Explain.

18. Monomeric saturated aliphatic carboxylic acids show carbonyl frequency near 1760 cm^{-1}, whereas, the saturated aliphatic ketones near 1720 cm^{-1}. Explain.

19. Explain why in *o*-hydroxybenzalehyde the frequency of the carbonyl compound is lower than *m*-hydroxybenzaldehyde.

20. An organic compound, molecular formula C_8H_8O, has a strong band near 1690 cm^{-1}. Assign the structure from the following structure:

(a) $C_6H_5CH_2CHO$

(b) $C_6H_5COCH_3$

(c) $C_6H_5OCH=CH_2$

(d) $HOC_6H_4CH=CH_2$

21. IR spectrum of acetone exhibited absorption at 1360 and 3000 cm^{-1}? Identify the C–H stretching and bending bands.

22. What modes of vibrations are active in IR absorption spectra and why? Taking nitro group as an example show the various types of vibrations which will be observed in the IR spectrum.

23. Indicate diagrammatically various modes of:

(a) Symmetric stretching

(b) Bending

(c) Symmetric stretching vibrations of CO_2

    Point out which of the above vibrations of CO_2 are IR active. Give reason?

24. Outline the use of IR spectroscopy in the study of hydrogen bonding?

25. Arrange the following with increasing carbonyl frequency. Explain the order:

    CH_3COCl CH_3CONH_2 CH_3COCH_3 $CH_3COOC_2H_5$

26. Explain the term vibrational frequency?

27. Write short notes on wagging and rocking vibrations?

28. How will you distinguish between acetic acid and acetone on the basis of their IR spectra when both show carbonyl stretching band at 1700–1750 cm^{-1}.

29. In both KBr and $CHCl_3$ solution o-nitrophenol exhibits its O–H stretching band at 3200 cm^{-1}, whereas the position varies in case of p-isomer in two media (3300 and 3520 cm^{-1}).

30. What is the necessary condition for a compound to be IR active?

31. How will you distinguish between cis and trans 1,2-dichloroethane on the basis of their IR spectra.

32. IR spectrum of dibenzalacetone in CS_2 at room temperature exhibits two separate peaks for C=O stretching vibrations. Explain.

33. Explain, why the 2-hydroxy-3-nitroacetophenone shows two carbonyl-stretching frequencies at 1692 and 1658 cm^{-1}?

34. Explain the observed variation in C=O stretching frequencies of the following compounds: Acid choride, ester and amide.

35. Explain, why in the substituted phenols the O–H stretching is at 3608 cm^{-1} in p-t-butyl phenol, and at 3605 and 3630 cm^{-1} in o-tert-butyl phenol, whereas the stretching frequency is at 3643 cm^{-1} in 2,6-di-tert-butyl phenol.

36. The carbonyl absorption of the carboxylate anion differs considerably from that of the parent acid. Explain.

37. How would you distinguish between the N–H stretching frequency of a primary amine and a secondary amine.

38. Where will cis 2-butene and trans 2-butene absorb in the IR spectra?

39. Explain, why trans 1-methylcyclohexane-1,2-diol has O–H infrared absorption bands at 3618 and 3597 cm^{-1}, whereas trans 1-isopropylcyclohexane-1,2-diol has a single peak at 3625 cm^{-1}?

40. How will you distinguish between axial and equatorial substituents in cyclohexanone on the basis of their IR spectra?

41. Explain, why axial 2-chloro-1-decalone absorbs at 1725 cm^{-1}, while equatorial isomer absorbs at 1750 cm^{-1} for carbonyl stretching?

42. How does the keto-enol tautomerism can be detected in an organic compound using IR spectroscopy?

43. Under ordinary conditions acetylacetone exists in tautomeric equilibrium. Comment on the IR spectral behavior of the compound.

44. How can you distinguish between 'gauche' and 'anti' conformers of 1,2-dichloroethane?

45. Explain, why cis-cinnamic acid absorbs at a higher frequency than the trans-isomer?

46. Why β-diketones frequently exist as mixtures of enolic and ketonic forms? Mention the various stretching and bending vibrations encountered in the enol form.

47. How will you determine the E and Z isomers on the basis of IR spectroscopy?

48. A compound in the vapour state absorbs for a particular bond at a higher wavenumber as compared to that when it is in the solid state. Explain.

49. The IR spectrum of two isomers A and B of molecular formula C_3H_6O indicates the absorption bands at 1650 and 1710 cm^{-1} respectively. Assign the structure of the two isomers.

50. Interpret the following spectra by assigning the bands for the given compounds:

(a) Benzoyl chloride

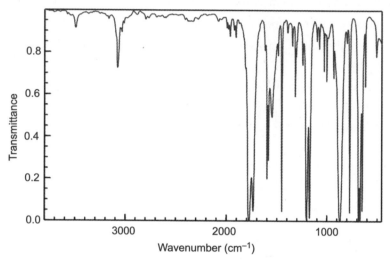

(b) Cyclopentanone

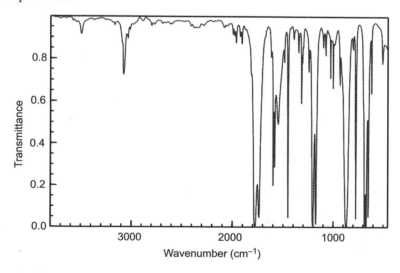

Chapter

4

Edward Mills Purcell

Felix Bloch

Proton Nuclear Magnetic Resonance (NMR) Spectroscopy

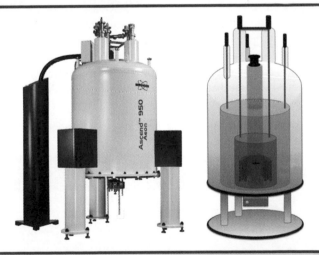

The founding fathers of nuclear magnetic resonance: Felix Bloch (1905–1983) and Edward Mills Purcell (1912–1997). Edward M. Purcell (August 30, 1912 – March 7, 1997) was an American physicist who shared the 1952 Nobel Prize for Physics, for his independent discovery of nuclear magnetic resonance in liquids and solids.

Chapter 4

Proton Nuclear Magnetic Resonance (NMR) Spectroscopy

4.1 INTRODUCTION

Nuclear magnetic resonance spectroscopy (NMR spectroscopy) deals with the magnetic properties of certain atomic nuclei. It involves the interaction between an oscillating magnetic field of electromagnetic radiation and the magnetic energy of a nuclei when placed in an applied magnetic field. It can be used to record differences in the magnetic properties of the various magnetic nuclei present, their nature, position and the atoms present in neighbouring groups in the molecule. It provides the information about the number of magnetically distinct atoms present in the molecule, so, it is more important than IR spectroscopy, which reveals the type of functional groups present in the molecule.

4.2 NUCLEAR SPIN STATES AND MAGNETIC MOMENT

In an applied magnetic field spin states are not of equivalent energy, because the nucleus is a charged particle, and any moving charge placed in a magnetic field generates its own magnetic field. The resulting spin-magnet has a magnetic moment (μ) proportional to the spin. This tiny magnet is referred as nuclear spin. Thus, a non-zero spin is always associated with a non-zero magnetic moment. It is the magnetic moment that allows the observation of NMR absorption spectra caused by transitions between nuclear spin levels.

Table 1. Isotopes of different elements, their spin number, magnetic moments and operating frequency.

Isotope	Occurrence in nature (%)	Spin number (1)	Magnetic moment (μ)	Operating frequency at 7 T (MHz)
^1H	99.984	1/2	2.79628	300.13
^2H	0.016	1	0.85739	46.07
^{12}C	98.9	0	–	–
^{13}C	1.1	1/2	0.70220	75.47
^{14}N	99.64	1	0.40358	21.68
^{15}N	0.37	1/2	– 0.28304	30.41
^{16}O	99.76	0	–	–
^{17}O	0.0317	5/2	– 1.8930	40.69
^{19}F	100	1/2	2.6273	282.40
^{31}P	100	1/2	1.1205	121.49

Overall spin of the nucleus is determined by the spin quantum number (I). If the number of both the protons and neutrons in a given nuclide are even, then I = 0 and zero nuclear magnetic moment, such nuclides are NMR inactive. Any atomic nucleus that posses even mass and odd atomic number or both mass as well as atomic number as odd, or odd mass and even atomic number, i.e., I ≠ 0, then the nuclei will be NMR active (**Table 1**).

4.3 THE ENERGY ABSORPTION AND RELAXATION PHENOMENA

4.3.1 The Resonance Phenomenon

The number of allowed spin states each nucleus with spin may adopt is quantized and is determined by its spin quantum number (I). The number I is a physical constant for each nucleus, and there are 2I+1 allowed spin states with integral differences ranging from +I to –I. A proton has the spin quantum number I = 1/2 and has two allowed spin states, i.e., 2(1/2)+1 = 2. For its nucleus the value of I may be +1/2 or –1/2. In the absence of an applied field, all the spin states of a given nucleus are of equivalent energy (degenerate), and in collection of atoms, all of the spin states should be almost equally populated with the same number of atoms having each of the allowed spins (**Figure 1a**).

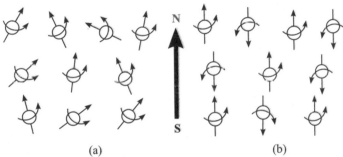

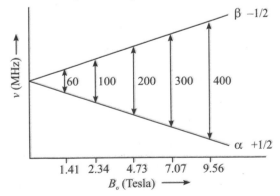

(a)   (b)

Figure 1. The orientation of nuclei (a) In the absence of applied magnetic field, (b) In the presence of applied magnetic field.

In the presence of an external magnetic field (B_o), its orientation will no longer be random (**Figure 1b**). The energies of two nuclear spin states become unequal and they may spontaneously "flip" from one energy state to the other. The energy required to induce flipping and obtain an NMR signal is just the energy difference between the two nuclear orientations and it depends upon the strength of the magnetic field B_o in which the nucleus is placed. The two spin states have the same energy when the external field is zero, but diverge as the field increases (**Figure 2**).

Figure 2. The spin states of a proton in the absence and presence of applied magnetic field.

The energy of an absorbed photon is $E = hv_o$, where, v_o is the resonance radiofrequency, should be equal to the Larmor precession frequency v_L of the nuclear magnetisation in the constant magnetic field B_o. Hence, a magnetic resonance absorption will only occur when $\Delta E = hv_o$, where $v_o = \gamma B_o/(2\pi)$. Such magnetic resonance frequencies correspond to the radio frequency range of the electromagnetic spectrum for magnetic fields upto ~20 T. It is this magnetic resonant absorption which is detected in NMR.

$$\Delta E = \gamma h B_o/2\pi \qquad \qquad ...(1)$$

Where, h is Planck's constant $(6.63 \times 10^{-27}$ erg sec)

$$\Delta E = hv_o \qquad \qquad ...(2)$$

On equating equations (1) and (2), v_o of the nuclear transition can be written as

$$v_o = \gamma B_o/2\pi \qquad \qquad ...(3)$$

Equation (1) is often referred to as the Larmor equation, and $\omega_o = 2\pi v_o$ is the angular *Larmor resonance frequency*.

$$\omega_o = \gamma B_o$$

The *gyromagnetic ratio* 'γ' is a constant for any particular type of nucleus and is directly proportional to the strength of the tiny nuclear magnet.

4.3.2 Relaxation Processes

The excited nucleus can undergo energy loss (or relaxation) leading to the relaxation of the molecule. If the relaxation rate is (a) fast, then saturation is reduced, (b) very fast, line-broadening is observed in the resultant NMR spectrum.

There are two major relaxation processes:

    (i) Spin-lattice relaxation (T_1)   (ii) Spin-spin relaxation (T_2)

4.3.2.1 Spin-lattice relaxation (T_1)

The excited nucleus can undergo relaxation by transferring ΔE to some electromagnetic vector present in the surrounding environment. This relaxation process is termed as spin lattice relaxation, where, lattice implies the entire framework or aggregate of neighbours. For example, the nearby solvent molecule, undergoing continuous vibrational and rotational motion, will have associated electrical and magnetic changes, which might be properly oriented and of the correct dimensions to absorb ΔE. This process keeps the excess of nuclei in the lower energy state, which is a necessary condition for nuclear magnetic resonance phenomenon.

    The relaxation time, T_1 (the average lifetime of nuclei in the higher energy state) is dependent on the gyromagnetic ratio of the nucleus and the mobility of the lattice. As mobility increases, the vibrational and rotational frequencies also increases, which leads to a higher probability of interaction of the excited nuclei with the lattice field component. However, at extremely high mobilities, this probability of interaction decreases. An efficient relaxation process involves small time, resulting in the broadening of the absorption peaks. Smaller the time of the excited state, greater is the line width. This mechanism is not effective in case of solids.

4.3.2.2 Spin-spin relaxation (T_2)

Spin-spin relaxation is the interaction between neighbouring nuclei with identical precessional frequencies but differing magnetic quantum states. In this process, the nuclei can exchange quantum states; a nucleus in the lower energy level will be excited, while the excited nucleus relaxes to the lower energy state. *There is no net change in the populations of the energy states, but the average lifetime of a nucleus in the excited state decreases resulting in dephasing (fanning out), line-broadening, and signal loss.*

    The nature of the peak depends on T_1 and T_2, if T_1 and T_2 are small than the broad peaks are obtained because of the short lifetime of the excited nucleus, whereas, if T_1 and T_2 are large the sharp peaks are obtained. The relationship between relaxation time and line broadening can be explained by uncertainty principle $\Delta E.\Delta t \approx h/2\pi$. Since $\Delta E = h\nu$, thus the product $\Delta\nu.\Delta t$ is constant. If Δt is large then $\Delta\nu$ is small (i.e., uncertainty in the measured frequency must be small so that there is very little spread in the frequency and line widths are narrow) and *vice versa*. For protons in the non-viscous solution sharp peaks are obtained because of the T_1 and T_2 relaxation times.

4.3.3 The NMR Spectrum

The NMR spectrum is the plot of absorption in the radio frequency region against chemical shift (δ). The NMR spectrum is generally determined on an approximately 10% solution of the test compound in deuterated solvent. Tetramethylsilane (TMS) is a widely used internal reference (or internal standard) compound due to its properties that includes; chemical inertness, low boiling, high solubility in most organic solvents and appears as a sharp singlet at 0.0 ppm in the NMR spectrum of the compound. TMS is insoluble in water or in D_2O, so sodium salt of 3-(trimethylsilyl)-propanesulfonic acid is used in place of TMS.

$$H_3C\!-\!\underset{\underset{CH_3}{|}}{\overset{\overset{CH_3}{|}}{Si}}\!-\!CH_3 \qquad\qquad H_3C\!-\!\underset{\underset{CH_3}{|}}{\overset{\overset{CH_3}{|}}{Si}}\!-\!CH_2CH_2CH_2SO_3Na$$

Tetramethylsilane (TMS)   Sodium 2,2-dimethyl-2-silapentane-5-sufonate (DSS)

4.4 INSTRUMENTATION IN NMR SPECTROSCOPY

A schematic diagram of the simple NMR instrument is shown in figure 3. The instrument consists of a strong magnet with a homogeneous field that can be varied continuously and precisely over a relatively narrow range. This is accomplished by the sweep generator. A radiofrequency transmitter, a radiofrequency receiver, a recorder, calibrator and integrator, a sample holder that positions the sample relative to the main magnetic field, the transmitter coil, and the receiver coil. The sample holder also spins the sample to increase the apparent homogeneity of the magnetic field.

By measuring frequency shifts from a reference marker, an accuracy of < 1 Hz can be achieved. The spectrum is presented as a series of peaks with the peak area being proportional to the number of protons they represent. Peak area is measured by an electronic integrator that traces a series of steps with height proportional to the peak area. The steps can be superimposed on the peaks. Proton counting is extremely useful. Peaks hidden under other peaks or in the baseline noise can be detected. Proton counting is often useful for determining sample purity and for quantitative analytical work.

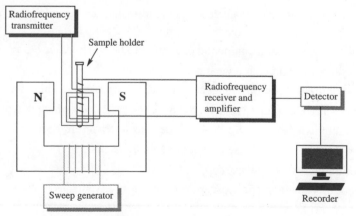

Figure 3. Schematic diagram of NMR spectrometer.

A routine sample on 300 MHz instrument is prepared by dissolving about 5–50 mg of the compound in about 0.4 mL of deuterated solvent in 5 mL O.D. glass tube. Fourier Transform can be used to obtain spectra of amounts as small as 5 mg.

4.4.1 Continuous Wave Spectroscopy

Nuclear magnetic resonance spectrometers used the technique of continuous-wave spectroscopy in early decades. This method involves a source of fixed frequency and the current (and hence magnetic field) is varied in an electromagnet to observe the resonant absorption signals. This technique is inefficient in comparison to Fourier technique, as it probes the NMR response at individual frequencies in succession. As the NMR signal is intrinsically weak, the observed spectra suffer from a poor signal-to-noise ratio. This can be mitigated by signal averaging i.e., adding the spectra from repeated measurements. While the NMR signal is constant between scans and so adds linearly, the random noise adds more slowly as the square-root of the number of spectra. Hence, the overall ratio of the signal to the noise increases as the square-root of the number of spectra measured.

4.4.2 Fourier Transform Spectroscopy

This method involved irradiating sample simultaneously with more than one frequency. A short strong pulse of radio frequency that possess all the frequencies distributed over whole of the ^{1}H range is applied to a set of nuclear spins that simultaneously excites all the single-quantum NMR transitions. In terms of the net magnetisation vector, this corresponds to tilting the magnetisation vector away from its equilibrium position (aligned along the external magnetic field). The out-of-equilibrium magnetisation vector precesses about the external magnetic field vector at the NMR frequency of the spins. This oscillating magnetisation vector induces a current in a nearby pickup coil, creating an electrical signal oscillating at the NMR frequency. This signal is known as the free induction decay (FID) and contains the vector-sum of the NMR responses from all the excited spins.

The advantage of FT-NMR is that the entire spectrum can be recorded, computerised and transformed in few seconds. Also, the spectrum of the sample can be obtained at (i) very low concentration; (ii) NMR studies on nuclei with low natural abundance and magnetic moments (e.g., ^{13}C, ^{15}N or ^{17}O).

═══════════ EXERCISE PROBLEMS 4.1 ═══════════

1. What is the basic principal of nuclear magnetic resonance spectroscopy, and what is its importance?
2. What are the condition for a compound or sample to be NMR active?
3. What is gyromagnetic ratio, and what is its significance?
4. What is spin-lattice relaxation? What are the different factors on which it depends?
5. What is spin-spin relaxation? What are the different factors on which it depends?
6. What is tetramethylsilane (TMS). Why it is used as internal reference?
7. What is the internal reference used for water soluble compounds?
8. What is continuous wave spectroscopy?
9. What is Fourier Transform spectroscopy? What are its advantages?

4.5 PARAMETERS OF NMR SPECTROSCOPY

4.5.1 Shielding and Deshielding of the Nucleus

When a molecule is kept in a magnetic field, the electrons around a nucleus start circulating, perpendicular to the applied magnetic field and induce a secondary magnetic field that can either oppose or reinforce the applied field depending on the position of the nucleus. If the induced field reinforce the applied field than the field experienced by the proton is greater than the applied field. Such a proton is said to be *deshielded* and this effect is termed as *deshielding effect*. On the other hand, If the induced field opposes the applied field than the field experienced by the proton diminishes and the proton is said to be *shielded* and this effect is termed as *shielding effect*.

4.5.2 Chemical Shift

In the presence of an applied field the shielding of the nucleus shifts the absorption position upfield (lower δ value), whereas, deshielding shifts the absorption position downfield (higher δ value). Such shifts in the NMR absorption positions are called chemical shift (**Figure 4**).

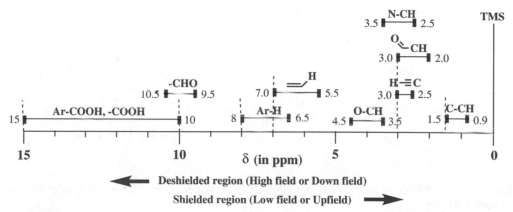

Figure 4. Schematic representation of the chemical shift range for different functional groups.

Chemical shift (δ) is usually expressed in parts per million (ppm) by frequency, because it is calculated from:

$$\delta_x = \frac{v_x - v_{TMS}}{v_o}$$

Where, δ_x is the chemical shift (in ppm), v_x and v_{TMS} are the frequencies in Hertz (Hz), for the signal x, the v_o is the operating frequency of the instrument in Megahertz (MHz). Thus, an NMR signal at 300 Hz from TMS at an applied frequency of 300 MHz has a chemical shift of:

$$\delta_x = \frac{300 \text{ Hz}}{300 \times 10^6 \text{Hz}} = 1 \times 10^{-6} \text{ ppm}$$

The frequency depends on the applied field, whereas, the chemical shift is independent of it. The most commonly used scale is δ. another scale is tau (τ). The relationship between τ and δ is given by:

$$\tau = 10 - \delta$$

4.5.3 Chemical Equivalence

In an organic compound protons with same environment absorb at same applied field, such protons are said to be equivalent. In other words, all of the protons found in chemically identical environments within a molecule are chemically equivalent, and they generally exhibit the same chemical shift (**Figure 5**).

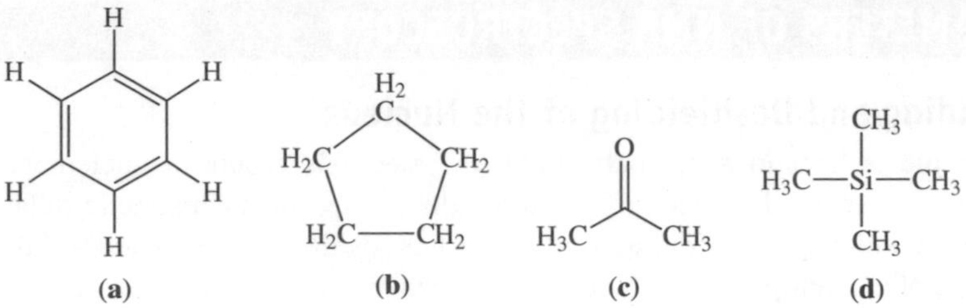

Figure 5. The molecules (a-d) show one NMR absorption peak, indicating all protons are equivalent.

In acetone, the two methyl groups are chemically equivalent due to the presence of a plane of symmetry and therefore all six hydrogen atoms of acetone appear at same δ value.

On the other hand, a molecule that has a set of protons which are present in different environment are chemically non-equivalent and they exhibits different absorption peaks in the NMR spectrum (**Figure 6**). For example, in 1,4-dimethyl benzene two different types of protons are present and therefore two different absorption peaks are observed in the NMR spectrum.

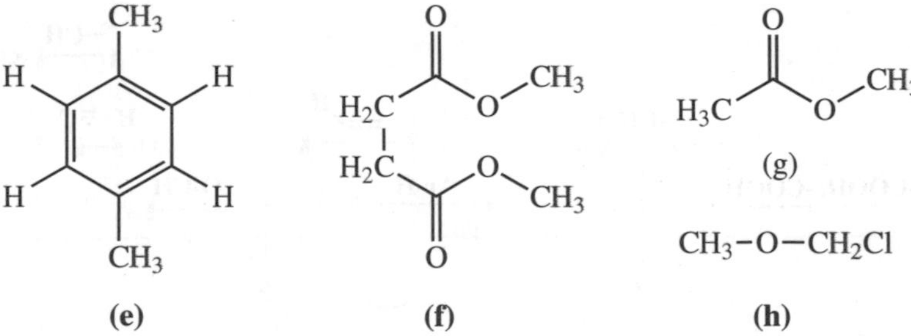

Figure 6. The molecules (e-h) show two NMR absorption peaks, indicating two sets of chemically equivalent protons.

Thus, valuable information can be obtained from the NMR spectrum about the number of peaks corresponding to the number of chemically distinct types of protons in the molecules.

4.5.4 Magnetic Equivalence

Two nuclei are said to be magnetically equivalent if they exhibit the same chemical shift and the same coupling constant to every other nucleus in the molecule. Often, chemically equivalent protons are also magnetically equivalent. However, in some instance protons are not magnetically equivalent. For example, in case of CH_3–CH_2–CH_3, there are 2 sets of protons (CH_3 and CH_2). The protons of a methyl group are magnetically equivalent since, as a consequence of the replacement of any of the three protons in methyl group gives the same product (**Figure 7**). In case, the replacement generates diastereomers, then the protons are magnetically non-equivalent.

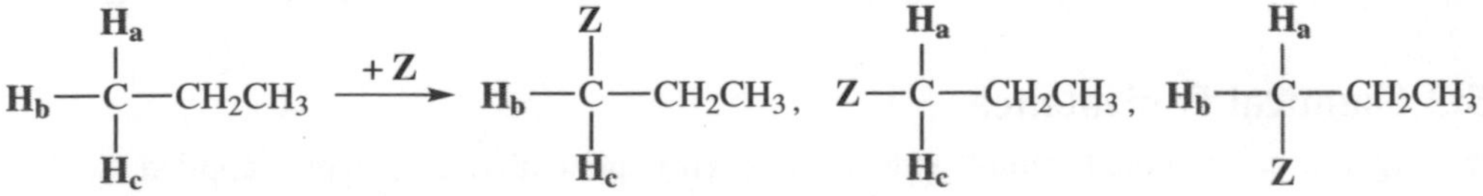

Figure 7. Magnetically equivalent protons.

4.5.4.1 Homotopic groups

These are atoms or groups on an atom that do not give a chiral molecule when one of the groups is replaced by another group. They have identical chemical shifts under all conditions (**Figure 8**). For example, the protons of the –CH_2 group of propane ($CH_3CH_2CH_3$) are magnetically equivalent which is called homotopic.

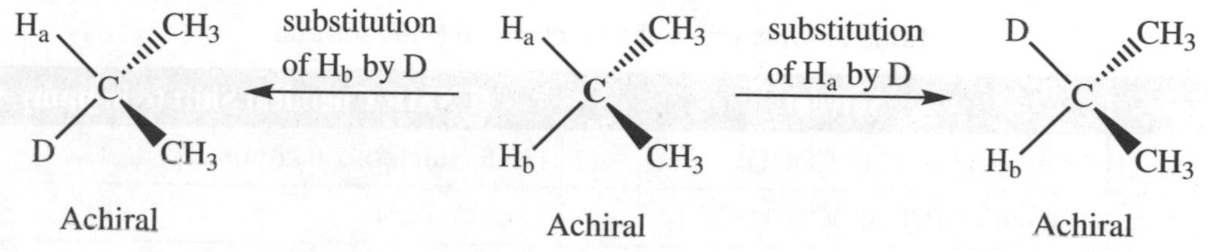

Figure 8. Substitution of homotopic groups generates achiral molecules.

4.5.4.2 Enantiotopic groups

In a molecule, when one of the atoms or groups is replaced by another group, a new stereocenter and a set of enantiomers is created, such atoms or groups of atom is known as enantiotopic group (**Figure 9**). They have identical chemical shifts in normal conditions but different chemical shift in chiral environments. For example, the protons of the –CH_2 group of CH_3CH_2Br are magnetically equivalent.

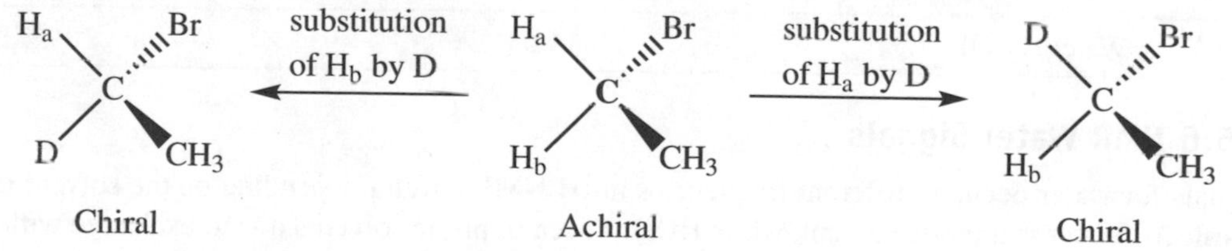

Figure 9. Substitution of enantiotopic groups generates chiral molecules.

4.5.4.3 Diastereotopic groups

In a molecule, when one of the atoms or groups is replaced by another group to give diastereomers such atoms or groups are called diastereotopic (**Figure 10**). Diastereotopic groups have different chemical shifts under all conditions. For example, in 1-bromoethylene two geminal protons are diastereotopic, as the replacement of one atom gives the Z-isomer while the other gives the E-isomer.

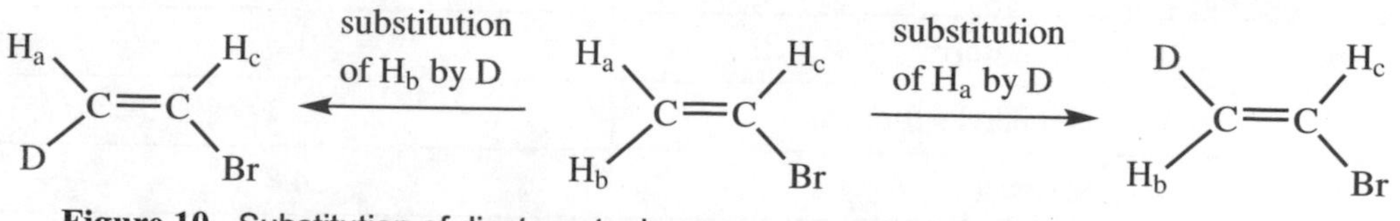

Figure 10. Substitution of diastereotopic groups generates diastereomeric molecules.

4.5.5 Nature of the Solvent

The ideal solvent should not contain protons in its structure. It should be inexpensive, low boiling, non-polar, and inert. Carbon tetrachloride (CCl_4) is ideal when the sample is sufficiently soluble in it. Carbon disulphide (CS_2) is also a useful solvent. The most widely used solvent is deuterated $CDCl_3$. The small sharp peak present from the $CHCl_3$ impurity present rarely interferes. The chemical shifts (δ) of solvent signals observed for ^1H NMR spectra are listed in the following (**Table 2**).

Table 2. Chemical shifts for common NMR solvents.

Entry	Solvent	^1H NMR Chemical Shift (δ in ppm)
1.	Acetic acid-d_4 (CD_3COOD)	11.65 (singlet), 2.04 (quintet)
2.	Acetone-d_6 (CD_3COCD_3)	2.05 (quintet)
3.	Acetonitrile-d_3 (CD_3CN)	1.94 (quintet)
4.	Benzene-d_6 (C_6D_6)	7.16 (singlet)
5.	Chloroform-d ($CDCl_3$)	7.26 (singlet)
6.	Dimethyl sulfoxide-d_6 (DMSO-d_6)	2.50 (quintet)
7.	Methanol-d_4 (CD_3OD)	4.87 (singlet), 3.31 (quintet)
8.	Methylene chloride-d_2 (CD_2Cl_2)	5.32 (triplet)
9.	Pyridine-d_5 (C_5D_5N)	7.58 (singlet), 7.22 (singlet)
10.	Water (D_2O)	4.63

4.5.6 NMR Water Signals

Signals for water occur at different frequencies in ^1H NMR spectra depending on the solvent used (**Table 3**). Water is a protic solvent, while HOD is seen in protic solvents due to exchange with the deuterium of solvent.

Table 3. Signals for water occur at different frequencies in ^1H NMR spectra.

Entry	Solvent	Chemical Shift of H_2O (or HOD, δ in ppm)
1.	Acetone	2.8
2.	Acetonitrile	2.1
3.	Benzene	0.4
4.	Chloroform	1.6
5.	Dimethyl sulfoxide	3.3
6.	Methanol	4.8
7.	Methylene chloride	1.5
8.	Pyridine	4.9
9.	Water (D_2O)	4.8

4.6 FACTORS AFFECTING CHEMICAL SHIFT

4.6.1 Inductive Effect

The chemical nature of an atom can influence its electron density through the polar effect. The inductive effect operates only along a chain of atoms and weakens with distance. When an electronegative atom is present on the atom adjacent to the proton, it withdraws electron density from the proton and the nucleus is therefore deshielded. However, when an electron-donating group (*viz.* alkyl group), is present on the atom adjacent to the proton, it increases the electron density on the proton resulting in the shielding of the nucleus. For example, in ^1H NMR of methyl halide (CH_3X) the chemical shift of the methyl protons increases in the order $I < Br < Cl < F$ from 2.2 to 4.3 ppm (**Table 4**). As the distance from the electronegative atom increases its deshielding effect on the proton decreases, and thus, the proton signal appears at a relatively higher field (lower δ value) as shown below.

	$CHCl_3$	CH_2Cl_2	CH_3Cl	$-CH_2Br$	$-(CH_2)_2Br$	$-(CH_2)_3\,Br$
Chemical shift (δ)	7.27	5.30	3.05	3.30	1.69	1.25

Also, if methyl is at terminal position on an alkyl chain, it appears at $\delta = 0.9$ ppm, but if any functional group is available, e.g., $-OCOR$ group nearby, on the β-carbon, than it has a minor deshielding effect (**Table 5**). The value given is $+ 0.4$ ppm; therefore, the CH_3 group appears at $\delta = 1.3$ ppm instead of 0.9 ppm. Similarly, the functional group at β-position of CH_2, CH influences the chemical shifts.

In case of benzene and its derivatives, the nature of the substituents may lead to shielding and desheilding of the protons (**Table 6**).

Table 4. Chemical shift value (δ) for CH_3, CH_2 and CH protons attached to X, where R = alkyl and Ar = aryl

Functional group (X)	Alkyl	CHO	COOH	CN	SH, SR	COAr	COR	CONH₂, CONHR, CONR₂	COOR, COOAr	F	Cl	Br	I	Ph, Ar	NH₂, NHR, NR₂	NHAr	NH-COR, NRCOR	N⁺R₃
CH₃X	0.9	2.2	2.1	2.0	2.1	2.5	2.1	2.0	2.0	4.3	3.0	2.6	2.2	2.3	2.1	2.5	2.9	3.3
R'CH₂X	1.3	2.2	2.3	2.5	2.4	2.8	2.4	2.0	2.1	–	3.4	3.3	3.1	2.6	2.5	3.1	3.3	3.4
R'R''CHX	1.5	2.4	2.6	2.7	2.5	2.9	2.5	2.1	2.2	–	4.0	4.1	4.2	2.9	2.9	–	3.5	3.5

Functional group (X)	NO₂	OH	OAr	OR	OCOR	OCOAr	—CHₐ—CH₂_b—O—	(C=C, CH₃)	(C=C)	(=·=·=·=)	(C=N—)	(≡)
CH₃X	4.3	3.4	3.7	3.3	3.6	3.9	1.3	1.7	2.0	1.8	2.0	2.0
R'CH₂X	4.4	3.6	3.9	3.3	4.1	4.2	a = 3.0	1.9	2.2	–	–	2.2
R'R''CHX	4.6	3.8	4.0	3.8	5.0	5.1	b = 3.5	2.6	2.3	–	–	–

(Table 4 Contd...)

Table 5. Influence of functional group X on chemical shift position of CH_3, CH_2 and CH protons at β-position to X.

X	C=C	CN	CO, CHO	COOH, COOR	OCOR, OCOAr	CONH₂	SH, SR	F	Cl	Br	I	Ph	OH, OR	OPh	NHCOR	NH₂, NHR, NR₂	NO₂
CH₃-C-X	0.1	0.5	0.3	0.2	0.4	0.25	0.45	0.2	0.6	0.8	1.0	0.35	0.3	0.4	0.1	0.1	0.6
CH₂-C-X	0.1	0.4	0.2	0.2	0.3	0.2	0.3	0.4	0.4	0.6	0.5	0.3	0.2	0.35	0.1	0.1	0.8
CH-C-X	0.1	0.4	0.2	0.2	0.3	0.2	0.2	0.1	0.0	0.25	0.4	0.3	0.2	0.3	0.1	0.1	0.8

Note: For β-shift add the above δ values given in Table 4.

Table 6. Shifts in the position of benzene protons ($\delta = 7.27$ ppm) caused by substituents.

X		Alkyl	Allyl	COOH, COOR	CN	CONH₂	COR	CHO	NH₃⁺	NH₂, NHR	N(CH₃)₂	NH-COR	NO₂	I	Br	Cl	OR	OH	OCOR	SO₃H, SO₂Cl, SO₂NH₂
	o	-0.15	0.2	0.8	0.3	0.5	0.6	0.7	0.4	-0.8	-0.5	0.4	0.1	0.3	0	0	-0.2	-0.4	0.2	0.4
	m	-0.1	0.2	0.15	0.3	0.2	0.3	0.2	0.2	-0.15	-0.2	-0.2	0.3	-0.2	0	0	-0.2	-0.4	-0.1	0.1
	p	-0.1	0.2	0.2	0.3	0.2	0.3	0.4	0.2	-0.4	-0.5	-0.3	0.4	-0.1	0	0	-0.2	-0.4	-0.2	0.1

4.6.2 Anisotropic Effect

The chemical shift in case of alkenes, alkynes, aromatic and carbonyl compounds cannot be explained on the basis of electronegativity. For example, the electronegativity of acetylenic proton is higher than that of ethylene protons, but the former absorbs at a lower δ value (δ 2.35 ppm) than the latter (δ 4.60 ppm). This can be explained by the anisotropic induced magnetic field effect which is result of a local induced magnetic field experienced by a nucleus resulting from circulating electrons. The induced magnetic field could be either paramagnetic when it is parallel to the applied field, or diamagnetic when it is opposed to it. These effects depend on the orientation of the molecule with respect to the applied field. Thus, a neighboring nucleus which may not be directly attached to the π-system may be shielded or desheilded depending on its orientation in space with respect to the π-system. In cyclohexane ring, the equatorial protons resonates 0.5 ppm higher than the axial protons this is attributed to anisotropic deshielding by σ-electrons in the βγ-bonds (**Figure 11**).

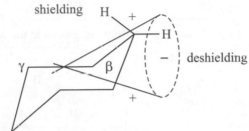

Figure 11. Anisotropic shielding and deshielding in cyclohexanes.

4.6.3 Hybridisation

4.6.3.1 Protons attached with sp^3 hybridised carbon

All the hydrogens attached to sp^3 hybridised carbon atoms have resonance in the range from 0 to 2 ppm, except in cases where the electronegative element or π-bonded groups are present. If the methyl group is attached to the sp^3 hybridised carbon than they resonate near 1 ppm, while methylene group hydrogens (attached to sp^3 carbons) appear near 1.2 to 1.4 ppm and tertiary hydrogens resonate near 1.8–2.0 ppm in the ^{1}H NMR spectrum. This can be explained on the basis that the axis of the C–C bond is in the axis of the deshielding cone in the former. The tertiary protons fall in the deshielding cone of three C–C bonds, secondary protons fall in the deshielding cone of two C–C bonds and primary protons fall in the deshielding cone of one C–C bond (**Table 7**).

Table 7. Effect of hybridisation on the chemical shifts.

Hybridisation	Functional group attached	Structure	Chemical shift (δ)
sp^3	Alkane	H|H−C−H|H	0.23
		CH$_3$|H−C−H|H	0.86

Hybridisation	Functional group attached	Structure	Chemical shift (δ)
		$H_3C-\underset{\underset{H}{|}}{\overset{\overset{CH_3}{|}}{C}}-H$	1.33
		$H_3C-\underset{\underset{CH_3}{|}}{\overset{\overset{CH_3}{|}}{C}}-H$	1.5
	Alkenes	$-C=C-\overset{|}{\underset{|}{C}}-H$	1.6 – 2.6
Attachment of electron withdrawing groups to the carbon			
	Carbonyl	$R-\overset{\overset{O}{\|}}{C}-\overset{|}{\underset{|}{C}}-H$ R = H, alkyl, OH	2.0 – 3
	Phenyl	$Ph-\overset{|}{\underset{|}{C}}-H$	2.2 – 2.7
	Nitrile	$N\equiv C-\overset{|}{\underset{|}{C}}-H$	1.7 – 2.8
sp^2	Alkene	$-\overset{|}{C}=\overset{|}{C}-H$	4.5 – 6.5
	Carbonyl	$\overset{\overset{O}{\|}}{-C}-H$	9.0 – 10
sp	Alkyne	$-C\equiv C-H$	1.6 – 3.0

4.6.3.2 Protons attached with sp^2 hybridised carbon

4.6.3.2.1 Alkenes

When an alkene molecule is placed in magnetic field it aligns itself perpendicular to the applied magnetic field and the circulation of the π-electrons generate an induced magnetic field in the same direction. Therefore, the induced magnetic field reinforces the applied field resulting in the deshielding of the olefinic protons. Thus, they absorb at higher δ value than expected from the electronegativity of olefinic carbon (**Table 8**, **Figure 12a**).

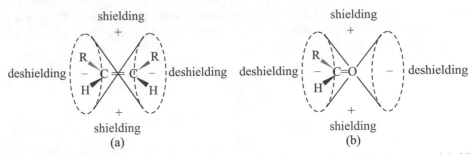

Figure 12. Schematic diagram of anisotropy caused by the presence of π-electrons (a) Alkenes; (b) Carbonyl compounds.

4.6.3.2.2 Carbonyl compounds

Aldehyde protons resonates near δ 9 to 10 ppm, which is higher than aromatic protons. This large chemical shift is attributed to anisotropy (**Table 8, Figure 12b**).

4.6.3.2.3 Aromatic compounds

Aromatic compounds consist of cyclic electron cloud delocalised cylindrically over the aromatic ring. When the aromatic ring is placed in a magnetic field and a magnetic field is applied perpendicular to the aromatic ring, circulation of π-electrons produces a ring current which induces a magnetic field perpendicular to the plane of the ring. This induced ring current opposes the applied field at the centre of the ring and reinforces the applied field outside the ring. Thus, the protons around the ring periphery absorbs at high δ value while the protons held above and below the plane of the aromatic ring resonates at comparatively lower δ value due to ring current effect (**Table 8, Figure 13**).

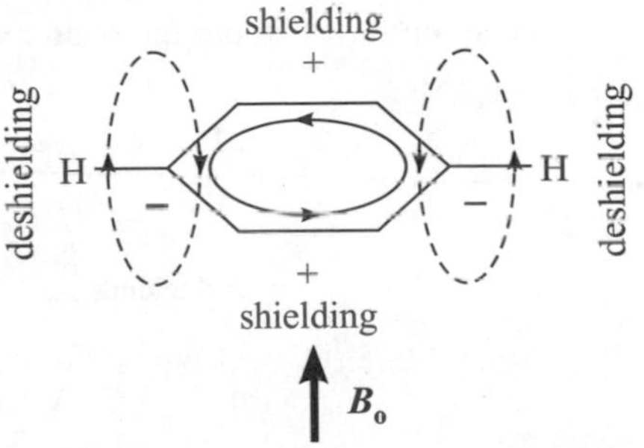

Figure 13. Shielding and deshielding zones in aromatic compounds.

The shielding and deshielding from aromatic ring current are stronger than that resulting from π-electrons of the olefinic bonds. In the molecule of toluene the methyl protons resonate at δ 2.34 ppm, whereas, a methyl group attached to acyclic conjugated alkene appears at δ 1.95 ppm. This can be attributed to the greater deshielding influence of the ring current in aromatic compounds (cyclically delocalised π-electrons) compared with the deshielding of conjugated alkene groups (having no cyclic delocalisation). Due to ring current 12 peripheral protons are deshielded (δ 8.9 ppm) and 6 internal protons shielded (δ −1.8 ppm) in [18]-annulene (**Figure 14**).

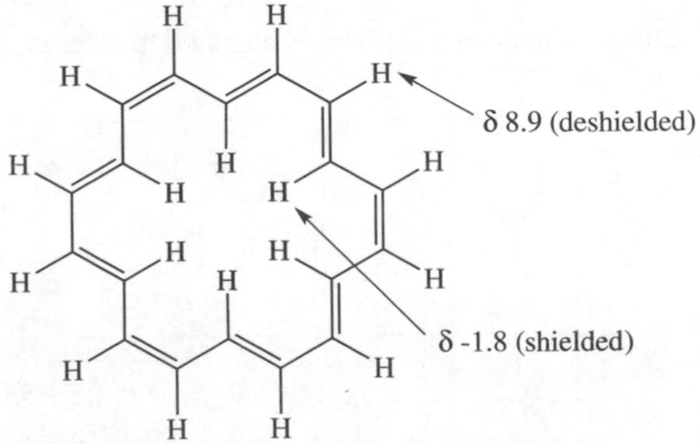

δ 8.9 (deshielded)

δ −1.8 (shielded)

Figure 14. Shielding and deshielding zones in [18]-annulene.

The methyl groups of dimethyl derivative of pyrene are in the shielding zone of the ring current and therefore appear at δ –4.2 ppm in the NMR spectrum. Thus, the above result indicates aromatic character in non-benzenoid system (**Figure 15**).

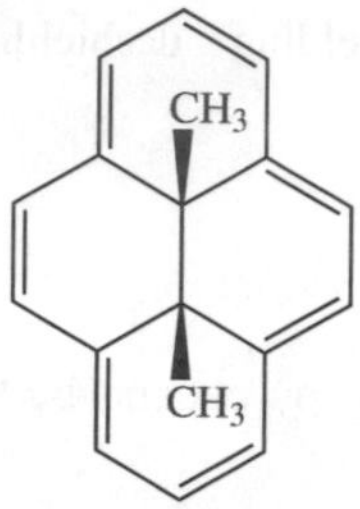

Figure 15. Shielding and deshielding region of dimethyl derivative of pyrene.

4.6.3.3 Protons attached with *sp* hybridised carbon

4.6.3.3.1 Alkynes

When an alkyne molecule is placed in magnetic field it aligns itself in such a manner that plane of its triple bond lies parallel to the applied magnetic field, this is because the π-electrons of the triple bond are symmetrical about the bond axis.

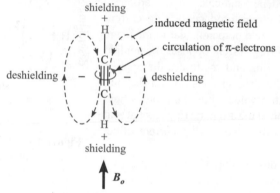

Figure 16. Shielding and deshielding region in acetylene.

The circulation of electrons about cylindrically π-electrons generate the induced magnetic field, which opposes the applied field in the region of alkyne protons. Thus, the acetylenic protons appear in the shielded region (low δ value) than expected from the electronegativity of acetylenic carbon (**Table 8, Figure 16**).

Table 8. Chemical shift values (δ) for H attached to different groups.

(i) Alkenes

Groups	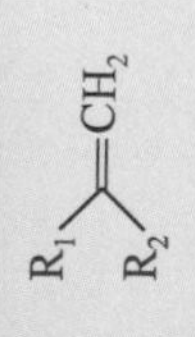 $R_1R_2C=CH_2$	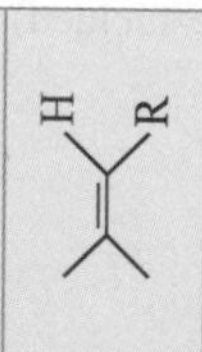 H/R	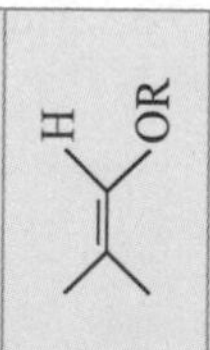 H/OR	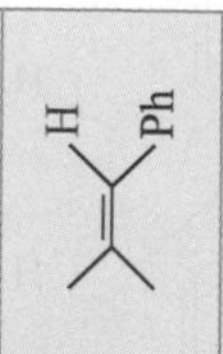 H/Ph	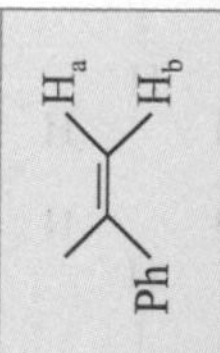 H_a/H_b Ph	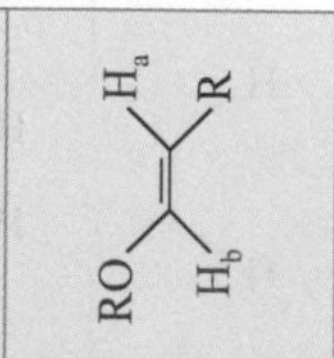 H_a/R RO H_b	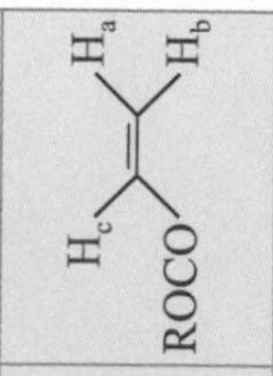 H_a/H_b H_c ROCO	Cycloalkene
Chemical shift (δ)	4.6	5.3	6.8	7.0	5.0 5.3	a = 4.5 – 5.0 b = 6.5 – 7.5	a = 4.5 b = 4.8 c = 7.2	5.6

(ii) Alkynes

Groups	$R-\!\!\equiv\!\!-H$	$Ph-\!\!\equiv\!\!-H$	$HO{\scriptstyle\diagdown}\!=\!\!\equiv\!\!-H$	$O{\scriptstyle\diagup}\!=\!\!\equiv\!\!-H$
Chemical shift (δ)	1.8	2.9	2.4	3.2

(iii) Enones

Groups					
Chemical shift (δ)	5.8	6.0	6.2	6.6	7.8

(iv) Miscellaneous

Groups						
Chemical shift (δ)	6.2	4.9	7.8	8.0	9.6 (R = Alkyl) 9.9 (R = Aryl)	α = 6.2 β = 4.5

Groups						
Chemical shift (δ)	7.27	α = 7.7 β = 7.5	α = 8.5 β = 7.0 γ = 7.4	X = O α = 7.4 β – 6.3	X = S α = 7.2 β = 7.1	X = N–H α – 6.5 β = 6.1

4.6.4 Hydrogen Bonding

The hydrogen bonding shifts the spectra to the higher δ values. Thus, stronger the hydrogen bond, more a proton will be deshielded (higher will be the δ values). For example, in carboxylic acids (δ = 10-12 ppm), inductive effect, resonance effect and hydrogen bonding combine to produce a large downfield shift. The degree of hydrogen bonding is solvent, temperature and concentration dependent. Thus, the chemical shifts of the acidic protons can also varies accordingly. The intermolecular and intramolecular H-bonding can be easily distinguished by using the NMR spectroscopy. Because the later will not show any shift in the absorption position on changing the concentration of the sample, whereas, the absorption position of the former sample is concentration dependent. For example, the absorption of hydroxyl proton of ethanol is shifted upfield on diluting the sample with non-polar solvent (CCl_4) due to the breaking of intermolecular hydrogen bond. Since, intramolecular hydrogen bonds are not broken on dilution, thus, these intramolecular hydrogen bonded protons show almost no change in their absorption on dilution.

4.6.5 Temperature

The resonance position of most signals is little affected by temperature, although OH, NH_2 and SH_2 protons resonate at a higher temperature, because the degree of hydrogen bonding is reduced.

================= **EXERCISE PROBLEMS 4.2** =================

1. What do you mean by shielding and deshielding of the nucleus in NMR spectroscopy?
2. What is chemical shift? What are the factors that affect chemical shift in ^{1}H NMR spectroscopy?
3. What is the relationship between Tau (τ) and delta (δ) units in NMR spectroscopy?
4. What do you mean by chemical equivalence and magnetic equivalence? Explain with example.
5. How inductive effect influences the chemical shift?
6. What is anisotropic effect?
7. Why the alkene protons resonates at higher value of chemical shift (δ) in comparison to alkynes?
8. Why aldehyde protons resonates near δ 9 to 10 ppm, which is higher than aromatic protons?
9. How intermolecular and intramolecular H-bonding can be distinguished using the NMR spectroscopy?

4.7 ORIGIN OF SIGNAL SPLITTING

4.7.1 Signal Coupling

An interaction in which the nuclear spins of adjacent atoms influence each other and lead to the splitting of NMR signals. The splitting of the spectral lines arises because of a coupling interaction between neighbouring protons. *Spin-spin splitting is independent of magnetic field strength, thus increasing the magnetic field strength will increase the chemical shift difference between two peaks in Hertz, but the coupling constant 'J' will not change.* There are different types of couplings which depend on the type of protons which are discussed below.

4.7.1.1 Geminal coupling

Protons attached to the same carbon atom are called geminal protons. These are separated by two bonds, and when they are non-equivalent they show spin-spin splitting. Geminal coupling constant (J_{gem}) is usually negative and increases algebraically on increasing the angle θ between the coupling protons.

4.7.1.2 Vicinal coupling

Protons attached to the adjacent carbon atoms are called vicinal protons. These are separated by three bonds. The coupling constant for two vicinal protons depend on the dihedral angle formed between the planes. The largest values for the coupling constants are obtained when the angle ϕ is $0°$ (eclipsed conformation) or $180°$ (*trans*-conformation); near $90°$ (gauche conformation), the coupling constant is nearly zero.

4.7.1.3 *Trans* coupling

In alkene groups, *trans* coupling ($J = 11$–19 Hz) is stronger than *cis* coupling ($J = 5$–14 Hz) (**Table 6**).

4.7.1.4 Aromatic coupling

Aromatic coupling depends on the position of the coupling protons with respect to each other, e.g., *ortho*, *meta* and *para* (**Table 9**).

Table 9. Proton spin-spin coupling constants of some common systems.

System	Coupling constant (J in Hz)
Geminal	10–18 (depending on the electronegativities of the attached groups)
Vicinal	Depends on the dihedral angle (HCCH dihedral angle)
	$^2J_{ab}$ 1–4
cis	$^3J_{ab}$ 5–14
trans	$^3J_{ab}$ 11–19
	$^3J_{ab}$ 4–10
	$^4J_{ab}$ 0–2
$\diagdown$C=CH$_a$–CH$_b$=C$\diagup$	$^3J_{ab}$ 10–13
	^3J_o 7–10 ^4J_m 2–3 ^5J_p 0–1
	J_{ae} 2–5 J_{ee} 2–5

System	Coupling constant (J in Hz)
	J_{ab} 3-5 J_{cd} 10-13

4.7.1.5 Long range coupling

In alkane systems (extending over more than three bonds) long range coupling is usually very small, but in olefins, acetylenes, aromatics, heteroaromatic and rigid systems, where the W-shaped zig-zag bonds are present, can be easily observed (**Table 10**).

Table 10. Proton spin-spin coupling constants of some long range coupling systems.

System	Coupling constant (J in Hz)
	$^4J_{ab}$ 1.8
	$^4J_{aa}$ 1–2
	$^4J_{ab}$ 7–8 $^4J_{cd}$ 0.2
	$^4J_{ab}$ 18
	$^4J_{ab}$ 3

System	Coupling constant (J in Hz)	
H_a (bicyclic structure with H_b)	$^4J_{ab}$	3–4
H_c (bicyclic structure with H_d)	$^4J_{cd}$	1–2
H_a (bicyclic structure with O and X, H_b)	$^4J_{ab}$	3
Long range coupling higher than ^4J:		
$CH_a-C=C-CH_b$	$^5J_{ab}$	1–2
$CH_a-C\equiv C-CH_b$	$^5J_{ab}$	2–3

4.7.2 Coupling Constant (J)

In NMR spectrum the distance between adjacent peaks in a multiplet is called the coupling constant (**Table 11**). It is measured in Hertz (Hz). It is a quantitative measure of the influence of the spin-spin coupling with adjacent nuclei. The value of 'J' depends upon the number, type and geometrical orientation of bonds separating the coupled nuclei. It is independent of the applied field, because splitting arises due to instantaneous spin states of neighbouring protons and not due to flipping of the spin states. It is a function of several factors, the most important of which are: (a) Number of bonds separating the nuclei, and (b) Stereochemistry of the molecule.

4.7.3 Multiplicity (Spin-spin Splitting/Spin-spin Coupling or n+1 Rule)

The number of lines observed in the NMR signal for a group of protons is not related to the number of protons in that group; but it is related to the number of protons in neighbouring groups (**Figure 17**). Spin-spin splitting is expected only between non-equivalent neighboring protons. Nuclei having the same chemical shift do not exhibit spin-splitting. Nuclei separated by three or fewer bonds (e.g., vicinal and geminal nuclei) will usually be spin-coupled and will show mutual spin-splitting of the resonance signals (same J values), provided they have different chemical shifts.

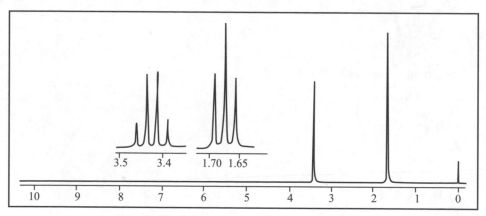

H_a gives a triplet for two protons (H_b & H_c)

H_b & H_c give a doublet for one proton (H_a)

H_d & H_e give a quartet for three equivalent protons (H_f, H_g & H_h)

H_f, H_g & H_h give a triplet for two equivalent protons (H_d & H_e)

Figure 17. Spin-spin splitting for different protons.

The splitting pattern of a given set of equivalent nuclei can be predicted by the n+1 rule, where, n is the number of neighbouring spin-coupled nuclei with the same (or very similar) J values. For example, in case of ethane (CH_3–CH_3) the protons of both methyl groups (all six protons) are equivalent. Hence, they do not split their signals and appear as a singlet.

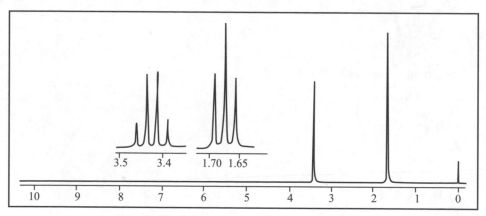

Figure 18. ^{1}H NMR spectrum of ethyl bromide.

But in case of ethyl bromide (CH_3–CH_2Br), three methyl protons are equivalent but they are non equivalent to methylene protons and *vice-versa*. Thus, three methyl protons spin-spin couple with methylene protons and split their signal into 3+1 = 4 peaks, i.e., methylene protons appear as a quartet. Similarly methylene protons split the methyl signal into 2+1 = 3 peaks. Thus, methyl protons will appear as triplet (**Figure 18**). If the proton responsible for spin-spin splitting are non-equivalent, than the number of peaks for a particular multiplet will be equal to (n+1) (n'+1) (n''+1), where, n, n' and n'' are the number of different kind of protons.

4.7.4 Relative Intensity of Lines in a Multiplet

The relative intensity of lines within a multiplet are given by numerical coefficient of the lines in the binomial expression $(1 + x)^n$ or by the use of Pascal's triangle, where, each term coefficient is the sum of the two terms diagonally above it (**Table 11**). For example, if, n = 1, the expression is $(1 + x) = 1 + x$, the lines of the doublet have relative intensities in 1 : 1 ratio. If, n = 2, then $(1 + x)^2 = 1 + 2x + x^2$. Thus the lines of the triplet have relative intensities in 1 : 2 : 1. If, n = 3, then $(1 + x)^3 = 1 + 3x^2 + 3x + x^3$. Thus, the lines of the quartet have relative intensities in 1 : 3 : 3 : 1. Hence, the splitting of a signal is due to the different environment of the absorbing proton not with respect to electrons but with respect to the nearby protons.

Table 11. Relative intensities of various multiplets.

Number of protons responsible for splitting (n)	Structure	Multiplet	Relative intensity
0	C–C–H (with C,C)	Singlet	1
1	H–C–H (C–C–H)	Doublet	1 : 1
2	H–C–H	Triplet	1 : 2 : 1
3	H–C–H	Quartet	1 : 3 : 3 : 1
4	H–C–C–H	Quintet (pentet)	1 : 4 : 6 : 4 : 1
5	H–C–C–H	Sextet	1 : 5 : 10 : 10 : 5 : 1
6	H–C–H	Septet	1 : 6 : 15 : 20 : 15 : 6 : 1

4.8 ANALYSIS OF THE NMR SPECTRUM

4.8.1 First Order Spectra

When the chemical shifts are large as compared to the coupling constant ($\Delta v/J^*$ is greater than about 10), δ and J values may be measured directly from the spectrum, such spectra are known as first order spectra. For example, 1,1-dibromoethane.

4.8.2 Non First Order Spectra

When, the chemical shifts of the coupled protons are of approximately the same magnitude as the coupling constants, the NMR spectra cannot be analyzed by the inspection and direct measurements in terms of the simple splitting rules. Such spectra are more complex and are known as non first order spectra. There are different types of coupled spectra which depends on the number and nature of interacting nuclei (**Table 12**).

Table 12. Different types of coupled systems.

S.No	Nuclei System	Type	Example
1.	Two	AX	Cinnamic acid
		AB	2-Methyl-3-hydroxymethyl furan
2.	Three	AX_2	1,1,2-Tricholoroethane
		A_2B	Pyrogallol
		AMX	Pyrrole-2-carboxylic acid
		ABX	Styrene
		ABC	Acrylonitrile; 1,2,4-Trichlorobenzene
3.	Four	A_2X_2	Phenyl acetate; β-Alanine.
		A_2B_2	2-Chloroethanol; 2-Phenoxyethanol
		AA′XX′	1,1-Diflouroethylene
		ABCD	Nicotinamide
4.	Five	A_2X_3	Ethyl acetate
		A_2B_3	Hexyl bromide

4.9 SIMPLIFICATION OF COMPLEX NMR SPECTRUM

When, the signals overlap with each other and there is not much difference between the chemical shift and coupling constant, more complex spectra are obtained. Thus, the complete analysis and extraction of useful information from NMR spectra becomes difficult. The important methods for simplifying a NMR spectrum to get maximum information are discussed below.

4.9.1 Increased Field Strength

When, the NMR spectra of an organic compound is analysed at low field (60 or 90 MHz) NMR instrument, the spectra is not very clear i.e., complicated. Since the chemical shift values of the protons of several groups are very similar, which results in the proton resonances in the same area of the spectrum, and often peaks overlap, so extensively that it is difficult to determine the individual peaks and their splitting. Such spectrum can be simplified using spectrometer operating at a higher frequency. For example, if a compound gives a complex NMR spectrum at 60 MHz, it will be improved by recording at higher frequency *viz* 300 MHz, although the degree of improvement depends on the particular chemical shift differences and the coupling constants involved (**Figure 19**). Also, the second order effects disappear at higher fields. As the strength of applied magnetic field increases, the chemical shift separation increases but the value of '*J*' remains constant, and thus $\Delta v/J$ ratio increases. Hence, the large proportion of the spectra becomes first order at high applied magnetic fields.

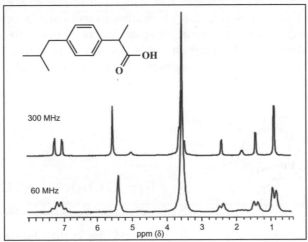

Figure 19. Effect of field strengths (60 MHz and 300 MHz) on the chemical shifts of protons of Ibuprofen.

4.9.2 Spin Decoupling or Double Resonance (Double Irradiation)

Spin decoupling of the complex spectra of the compound can be reduced to the first order spectra by the removal of some of the interactions provided the chemical shift positions for the coupling multiplets should not be closer than ≈1 ppm. A second strong field (H_2) is applied to sense line or multiplet that is coupled to another and examining the resulting spectrum. When a multiplet is irradiated strongly it becomes saturated and undergoes rapid transitions between its two spin states, and hence is no longer affected by any adjacent nuclei. Thus, the spectrum simplifies because the coupling between that nucleus or other group of nuclei has been removed.

$$\Delta t . \Delta \upsilon \approx 1/2\pi$$

For pair of coupled protons the time averaged needed to resolve the two lines of a doublet (Δt) is related to the separation between the lines, i.e., coupling constant. Since, simultaneous two radiofrequency sources are used the technique is called double resonance or double irradiation. Since, the nuclear spins during the process are less coupled than before, therefore it is called as spin decoupling.

4.9.3 Deuteration-deuterium Exchange and Deuterium Labelling

In compounds with labile protons such as OH, NH_2, SH, active methylene flanked by carbonyl groups can exchange protons with D_2O, such an exchange is known as deuterium exchange. Such protons can be identified by shaking the sample solution with D_2O and then re-running the spectrum. The acidic protons will get exchanged with deuterium atoms, which disappear from the spectrum, and is replaced by a singlet due to HOD ($\delta = 4.8$ ppm). Hydrogen and deuterium nuclei have broad difference in their magnetic properties. Thus, it is possible to distinguish between them by NMR spectroscopy. For example, in ROH, when D_2O is added it forms ROD and the signals due to OH disappears and a new signal for HOD appears. Similarly, if D_2O is added to RCOOH then due to rapid exchange it becomes RCOOD.

4.9.4 Shifts Reagent

The shifts reagents are generally the complexes of the Lewis acids of paramagnetic rare earth metals from the lanthanide series. The most commonly used shifts reagents are the complexes of europium (III) and praseodymium (III) (**Figure 20**). Their reactivity as shifts reagents depend entirely on their ability to bind basic molecules (heteroatoms with a lone pair of electrons, e.g., oxygen, nitrogen, etc.) as labile ligands. These are the reagents, which allow a rapid and relatively inexpensive means of resolving overlapping multiplets in some complex spectra. When such metal complexes are added to the compound for which the spectrum is to be determined, profound shifts in the resonance positions of the various groups of protons are observed. The direction of shift (upfield or downfield) depends on the nature of metal used.

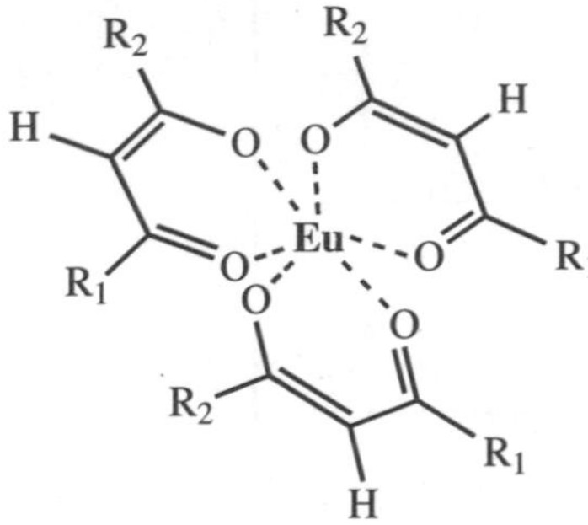

$$R_1 = R_2 = -C(CH_3)_3 \equiv Eu(DPM)_3$$
$$R_1 = -C(CH_3)_3, R_2 = -C_3F_7 \equiv Eu(FOD)_3$$
$$R_1 = R_2 = -C_2F_5 \equiv Eu(FHD)_3$$

Figure 20. Different Europium based shifts reagents.

Complexes of europium, erbium and ytterbium, shifts resonances to larger δ values, while complexes of cerium, praseodymium, neodymium, samarium, terbium and holmium, generally shifts the resonances to smaller δ values. The position of a given peak is related to the stability of the complex formed, the amount of shifts reagent added, and the location of the observed nucleus relative to the lanthanide ion. The advantage of using such reagents is that shifts similar to those observed at higher δ values can be induced without the purchase of high field strength instrument. The lanthanide complex should be soluble in common NMR solvents for wide applicability, and those most frequently used are complexes with two enolic β-diketones, dipivaloylmethane (DPM) and heptaflurodimethyloctanedione (FOD). Decafluoroheptanedione (FHD) has no proton and is soluble in CCl_4.

EXERCISE PROBLEMS 4.3

1. What is spin-spin splitting?

2. What is coupling constant? What are its units and how it is affected by changing magnetic field strength?

3. What is 'geminal' and 'vicinal' coupling? Explain with suitable examples.

4. What is first order spectra? Give example.

5. What is non first order spectra? Give example. How it can be simplified using different techniques in NMR spectroscopy?

6. What are shifts reagents? Give examples.

4.10 APPLICATIONS OF ^{1}H NMR SPECTROSCOPY

4.10.1 Detection of Aromaticity

It is possible to determine experimentally whether a compound has a close ring of electrons or not. If on application of the magnetic field the protons attached to the ring are shifted downfield from the normal olefinic region than molecule will be aromatic. In addition, if the compound possess protons above or within the ring than these gets shifted to upfield.

4.10.2 Detection of Rate Constant by Environmental Exchange and Line Broadening

The hydroxyl proton in the commercial sample of ethanol appears as a singlet even though it is associated with several magnetic environments in the medium. This occurs when the rate constant for the exchange of the protons from one environment to another is greater than the frequency difference of the proton resonances in the separate environment. This means that the rate constant for the exchange of protons will affect resolution of proton signals. This may be explained as follows.

  (i) When the rate constant for the exchange of protons is very slow, the protons appear as separate signals.

 (ii) The rate constant for the exchange of protons is very fast, the protons appear as singlet.

(iii) The rate constant for the exchange of protons is comparable to the frequency difference, broad lines are observed.

4.10.3 Distinction between *cis trans* Isomers and Conformers

Proton NMR can be used to distinguish the *cis-trans* isomers because the protons show a difference in the chemical shift as well as coupling constant for example in alkene groups, *trans* coupling ($J = 11$–19 Hz) is stronger than *cis* coupling ($J = 5$–14 Hz).

4.10.4 Detection of partial double bond character

The ^{1}H NMR spectrum of N,N-dimethylformamide recorded around room temperature, shows two signals for methyl groups. However, it is expected that the two methyl groups would be in magnetically equivalent environment due to the free rotation along C–N bond to the carbonyl group.

N,N-Dimethylformamide is represented by two resonance forms (**Figure 20a**), and as a result of conjugation between the carbonyl group and the non-bonding pair on nitrogen the double bond character of the C–N bond sufficiently increases and restricts the free rotation at room temperature. Thus, one methyl group is *cis* to oxygen, the other is *trans*, and the anisotropy of the carbonyl group is sufficient to influence the chemical shift position for the *cis* group. When the sample is heated upto 130 °C than the spectrum of N,N-dimethylformamide shows only one signal for the methyl group (**Figure 20b**). This can be attributed to the fact that at elevated temperature rotation around the C–N bond is so rapid that each methyl group experiences the same time averaged environment and appears as a singlet. Similar effects are found in fluxional molecules and in structures where steric effects can intervene; for example, cyclic structures like cyclohexane ring-flips, bridged structures as in α-pinene, spirans etc.

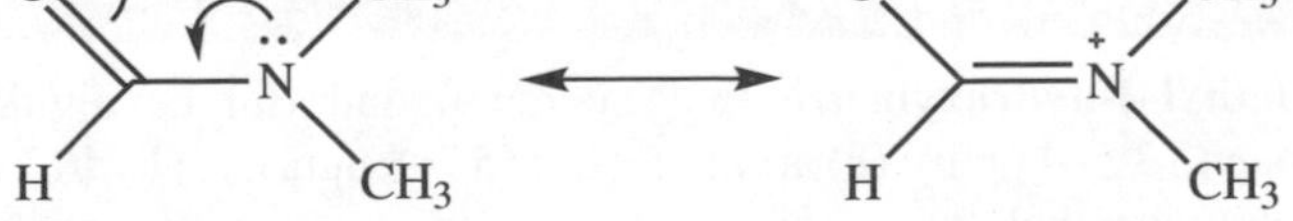

Figure 20a. Resonating structure of N,N-dimethyl formamide.

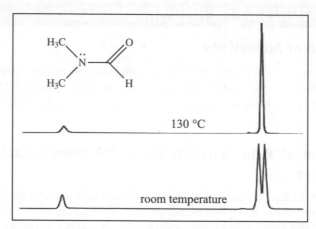

Figure 20b. Effect of temperature on the ¹H NMR spectra of *N,N*-dimethyl formamide.

4.10.5 Quantitative Analysis in Keto-enol Tautomerism

¹H NMR spectroscopy has been used to study the keto-enol equillibria in β-diketone and β-ketoesters, also to determine the relative amounts of tautomers in a tautomeric mixture. For slow tautomerism, the various signals in keto and enol forms occur independently and are well separated. It is thus possible to determine the keto-enol ratio and also the equilibrium constant by comparing the relative intensities of keto ($-CH_2$) and enol ($=CH$) signals. For example, the spectrum of acetylacetone exhibits clearly the absorption due to keto and enol forms (**Figure 21**).

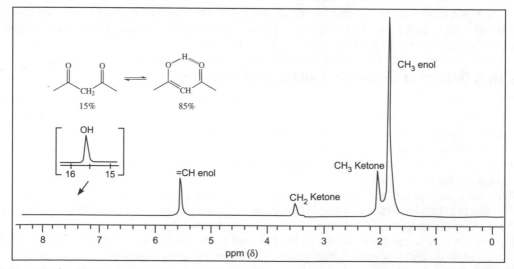

Figure 21. ¹H NMR spectra of acetylacetone for the quantitative analysis of keto-enol tautomers. The keto-enol ratio was measured by integration of the CH_3 peaks.

4.11 INTERPRETATION OF SPECTRA

Compound A: Methyl-4-hydroxybenzoate: The compound can be divided into three parts (a) methyl ester group (δ 3.5–4 ppm); (b) aromatic ring (δ 7–8 ppm); (c) hydroxy group (δ 6–7 ppm) from ¹H NMR spectra given below:

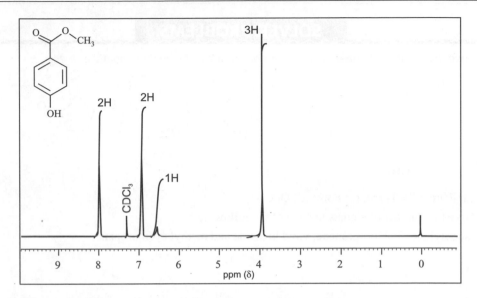

Using the NMR spectra more specific and integral values can be assigned as ester group: 3.79 (s, 3H, CH$_3$); (ii) aromatic ring: 6.82 (d, 2H, H-3, H-5), 7.90 (d, H-2, H-6); (iii) hydroxy group: 6.52 (s, 1H, OH). Finally take into account the spin-spin coupling. The hydroxy group will appear as a singlet, since it does not have any neighbouring proton, the four aromatic protons will appear as two doublets for two protons each, since it is *p*-disubstituted ring. The ester methyl will appear as a singlet because of the absence of any neighbouring proton.

Compound B: *p*-Hydroxyacetophenone: The compound can be divided into three parts (a) methyl ketonic group (δ 3.5–4.0 ppm); (b) aromatic ring (δ 7–8 ppm); (c) hydroxy group (δ 5–7.5 ppm) from Figure 4. Using the NMR spectra more specific and integral values can be assigned to alkyl group: δ 2.42, (s, 3H, CH$_3$); (ii) aromatic ring: δ 6.94, (d, *J* = 8.4 Hz, 2H, H-3 and H-5), 7.92, (d, 2H, H-2 and H-6); (iii) hydroxy group: δ 7.14, (s, 1H, OH). Finally take into account the spin-spin coupling. The hydroxy will appear as a singlet as it does not have any neighbouring proton. The aromatic protons will appear as two doublets for two protons each, since it is *p*-disubstituted ring. The methyl and hydroxy groups will appear as singlets as they do not have neighbouring protons.

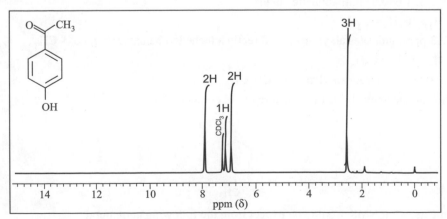

SOLVED PROBLEMS

Q1. Give the number of non-equivalent protons in the following compounds:

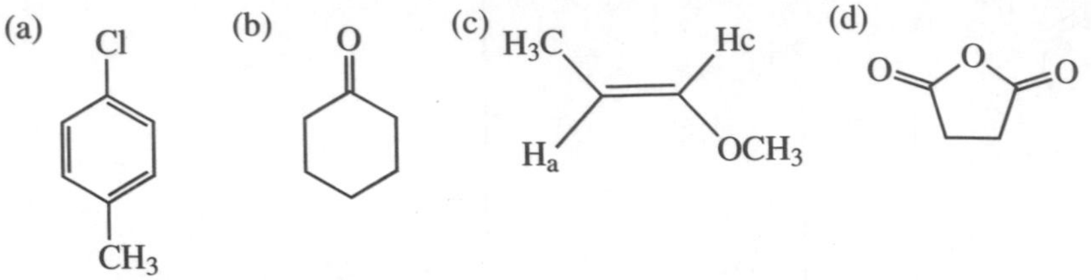

(a)   (b)   (c)   (d)

Sol. (a) Three; (b) Three; (c) Four; (d) One

Q2. Give the number of signals in each of the following:

Isobutylene, 2-chloropropene, vinyl bromide and methyl cyclopropane

Sol.

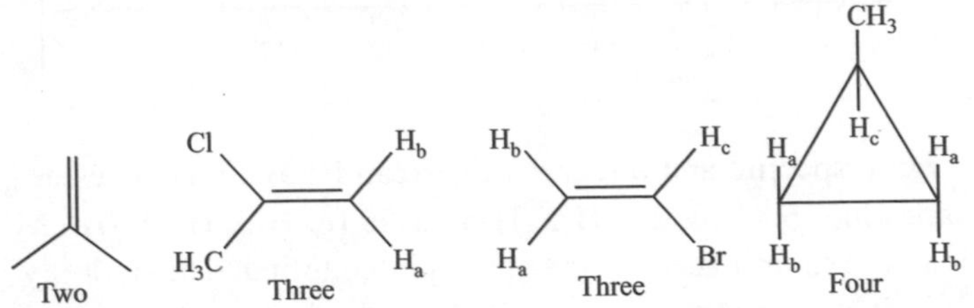

Two   Three   Three   Four

Q3. Determine the structure of the compound with molecular formula C_9H_{12} showing ^1H NMR signals at δ 7.1, 2.3, 1.6 and 0.8 ppm.

Sol. Calculate the degree of unsaturation for C_9H_{12}

$DBE = 9 + 1 - 12/2 = 10 - 6 = 4$

Thus, the given compound has four degree of unsaturation.

Now consider the given δ values: 7.1, 2.3, 1.6 and 0.8 ppm

From the given data it could be inferred that the compound consist of four types of protons.

δ = 7.1 ppm, Aromatic group may be present which can satisfy the degree of unsaturation also. As in benzene ring three double bonds are present and one for cyclic ring. Therefore C_6H_5 could be one unit.

δ = 2.3, 1.6, 0.8 ppm, aliphatic chain

$C_9H_{12} - C_6H_4 = C_3H_8$

δ 2.3 ppm, indicates alkyl group is directly attached to benzene ring, i.e., $-CH_3$

$C_3H_8 - CH_3 = C_2H_5$

1.6, 0.8 indicates an alkyl group, $-CH_2CH_3$

Thus on the above basis the structure of the compound is:

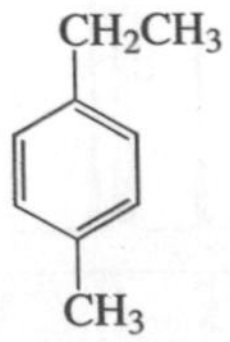

Q4. Indicate, what NMR would you expect from the following compound.

(i) CH_3COCH_3 (ii) $CH_3CH_2CH_3$ (iii) $CH_3CH_2CH_2CH_3$

Sol. In CH_3COCH_3, there are one type of protons (equivalent) and thus one signal as singlet is expected.

(ii) $^aCH_3{}^bCH_2{}^aCH_3$, there are two types of proton and thus two signal are expected. The two signal are (a) 6H, triplet and (ii) 2H as septet.

(iii) $^aCH_3{}^bCH_2{}^bCH_2{}^aCH_3$, there are two type of protons and thus two signal are expected. The two signals are (a) 6H, triplet and (b) 4H as quartet.

Q5. An organic compound with molecular formula C_5H_{12} shows two signals in its NMR spectrum, a triplet at δ 0.90 ppm for 6H, and a multiplet at δ 1.24 ppm for 6H.

Sol. The degree of unsaturation for C_5H_{12}

Thus,   DBE $= 5 + 1 - 12/2 = 6 - 6 = 0$

Thus, the given compound is saturated.

Now consider the given δ values: 0.9 for 6H, and 1.24 ppm for 6H

From the given data it could be inferred that the compound consist of two types of protons.

(a) $\delta = 0.9$ ppm as triplet for 6H indicates that two methyl group are present in neighbouring position of CH_2 group that gives the signal at 1.24 ppm for 6H.

Thus, on the above basis the structure of compound is:

$$CH_3CH_2CH_2CH_2CH_3$$

Q6. Explain, why OH and NH_2 appears as broad signals?

Sol. This could be explained on the basis that the relaxation time of protons are often shortened by chemical exchange processes and other nuclear phenomena.

Q7. How will you distinguish between chlorobenzene and 1,2-dichloroethane?

Sol. Chlorobenzene shows a singlet for 5H at δ 7.3 ppm, while 1,2-dichloroethane a singlet due to a set of four equivalent protons which appears at high δ value nearly 2.5 ppm.

Q8. How will you distinguish ethylacetate and methylpropionate by NMR?

Sol. Ethylacetate ($^aCH_3^bCH_2COO^cCH_3$) has three types of protons. Three protons are attached to methylene group will appear as a triplet at nearly δ 0.9 ppm and methylene group attached directly to oxygen will appear as a quartet nearly δ 4.2 ppm and the methyl group attached to carbonyl group will appear as a singlet at δ 2.1 ppm.

Methylpropionate ($^aCH_3^bCH_2^cCH_2COO^dCH_3$) has four types of protons. Three protons attached to methylene group will appear as a triplet at nearly δ 0.9 ppm and methylene group attached directly to oxygen will appear as a triplet nearly δ 4.2 ppm, the methylene sandwiched between a methyl and methylene group will appear as a multiplet at δ 2.5 ppm and the methyl group attached to carbonyl group will appear as a singlet at δ 2.1 ppm.

Q9. Predict the structure of a compound with molecular formula $C_9H_{11}Br$, showed the following ^{1}H NMR data δ 2.25 ppm, 2H, multiplet; δ 2.75 ppm, 2H, triplet; δ 3.38 ppm, 2H, triplet; δ 7.22 ppm, 5H, singlet.

Sol. Calculate the degree of unsaturation for $C_9H_{11}Br$

DBE $= 9 + 1 - (11 + 1)/2 = 10 - 6 = 4$

Thus, the given compound has four degree of unsaturation.

Now consider the given δ values: 7.22, 3.38, 2.75 and 2.25 ppm

From the given data it could be inferred that the compound consist of four types of protons.

$\delta = 7.22$ ppm, singlet for 5H indicates than an aromatic nucleus is present which can satisfy the degree of unsaturation also. As in benzene ring three double bonds are present and one for cyclic ring. Therefore C_6H_5 could be one unit.

$\delta = 3.38, 2.75$ and 2.25 ppm aliphatic chain

$$C_9H_{11} - C_6H_5 = C_3H_6$$

A two proton multiplet at δ 2.25 and a two proton triplet at δ 2.75 ppm and two proton triplet at δ 3.38 ppm shows the following combination.

$$-^aCH_2{}^bCH_2{}^cCH_2Br$$

$-^aCH_2$ at δ 2.25 ppm as 2H triplet downfield indicates that it is directly attached to aromatic ring

$-^bCH_2$ protons are under the influence of four protons (two on either side) and thus appears as multiplet.

$-^aCH_2$ at δ 3.38 ppm as 2H triplet downfield indicates that it is directly attached to halogen.

Thus, on the above basis the structure of compound is:

Q10. What will be the multiplicity of each kind of proton in the following molecules.

Sol. (i) This compound has two sets of equivalent protons which appears as singlet, *i.e*, one set of 9 protons and other set of 2 protons.

(ii) In this compound the two protons are present in different environment due to which both the protons can undergo spin-spin coupling and will appear as doublets.

(iii) In this compound both the protons will give rise to a doublet due to spin-spin coupling.

UNSOLVED PROBLEMS

1. How many proton signals would you expect to possibly find in the ^{1}H NMR spectrum of 2-chloropentane?
   (a) 6   (b) 7   (c) 8   (d) 9   (e) More than 9

2. How many chemically non-equivalent protons are there in 2,2-dimethylbutane?
   (a) 2   (b) 3   (c) 4   (d) 5

3. What is the multiplicity (spin-spin splitting) of the protons of 1,2-dichloroethane?
   (a) one singlet   (b) two singlets   (c) one doublet (d) one triplet

4. Which isomer of C_4H_9Br has only one peak in its ^{1}H NMR spectrum having the chemical shift δ 1.8 ppm?
   (a) 1-Bromobutane   (b) 2-Bromobutane
   (c) 1-Bromo-2-methylpropane   (d) 2-Bromo-2-methylpropane

5. Which isomer of formula $C_4H_6Cl_4$ has two signals at δ 3.9 (doublet, 4H) and 4.6 ppm (triplet, 2H) in its ^{1}H NMR spectrum?

   (i)   (ii)   (iii)   (iv)

6. Which of the following statements is false?
   (a) Splitting of the hydroxyl proton of an alcohol is not usually observed.
   (b) Alcohol protons shift to lower fields in more concentrated solutions.
   (c) Addition of D_2O to alcohol will result in an increased intensity of the hydroxyl proton signal.
   (d) The chemical shift of the hydroxyl proton depends on solvent, temperature, and concentration of the solution.

7. What do you expect to observe in the ^{1}H NMR spectrum of chloroethane CH_3CH_2Cl?
   (a) A triplet and a quartet   (b) Two doublets
   (c) A doublet and a triplet   (d) A doublet and a quartet

8. How many signals do you expect to see in the ^{1}H NMR spectra of 2-bromopropane $[(CH_3)_2CHBr]$ and 1-Bromopropane $(CH_3CH_2CH_2Br)$?
   (a) 1-Bromopropane = 2; 2-Bromopropane = 2
   (b) 1-Bromopropane = 3; 2-Bromopropane = 2
   (c) 1-Bromopropane = 2; 2-Bromopropane = 3
   (d) 1-Bromopropane = 3; 2-Bromopropane = 3

9. A ^{1}H NMR spectrum of compound C contains a singlet, a triplet and a quartet. Which of the following compounds might be C?
   (a) $CH_3CHClCHClCH_3$ (b) $CH_3CH_2CH_2CHCl_2$
   (c) $CH_3CH_2CHClCHCl_2$ (d) $CH_3CCl_2CH_2CH_3$

10. In the IR spectrum of a compound X, there is a strong absorption at 1718 cm^{-1} in addition to bands at 2978 and 2940 cm^{-1} and bands below 1500 cm^{-1}. The ^{1}H NMR spectrum contains two signals: A quartet and a triplet with relative integrals of 2 : 3. Of the following compounds, which is most likely to be X?

(a) $CH_3CH_2CO_2H$ (b) $CH_3CH_2COCH_2CH_3$

(c) CH_3CH_2OH (d) $CH_3CH_2OCH_2CH_3$

11. Vicinal coupling is:

(a) Coupling between ^1H nuclei in an alkane.

(b) Coupling between ^1H nuclei attached to the same C atom.

(c) Coupling between ^1H nuclei attached to adjacent C atoms.

(d) Coupling between ^1H nuclei in an alkene.

12. The observed chemical shift of a proton is 300 Hz from TMS and operating frequency of the spectrometer is 100 MHz. Calculate the chemical shifts in term of δ (ppm).

13. Acetylinic protons are more shielded than ethylenic protons, although the former are attached to the more electronegative sp-carbon. Explain.

14. Explain, why the aldehydic protons appears much downfield in the ^1H NMR spectrum.

15. Explain, why the aromatic protons appears much downfield in the ^1H NMR spectrum than ethylenic protons.

16. Explain why broad signals are often observed in the ^1H NMR spectra associated with OH and NH resonances.

17. How many different types of protons are present in the molecule of allyl bromide?

18. Why the ^1H NMR spectrum of dimethyl formamide shows two signals at δ 2.04 and 3.0 ppm at room temperature, but a sharp signal at higher temperature 130 °C?

19. How will you distinguish between o-hydroxy benzaldehyde and p-hydroxy benzaldehyde on the basis of ^1H NMR spectroscopy?

20. Why in the ^1H NMR spectrum of pure anhydrous ethanol the OH proton appears as a triplets, while in the impure sample it appears as a singlet.

21. How will you distinguish cis and $trans$-stilbine using ^1H NMR spectroscopy?

22. How will you distinguish between equatorial and axial protons in cyclohexane?

23. Predict the approximate chemical shift position for each of the different hydrogens in the ^1H NMR spectrum of this compound. Also predict the multiplicity and integral for each of the signals of each of the signals in the ^1H NMR spectrum of the compound.

$$\underset{\displaystyle CH_2CHCOCH_2CH_3}{\overset{\displaystyle Cl \qquad O}{\overset{\displaystyle | \qquad ||}{}}}$$

24. Determine the structure of the unknown compound with the molecular formula $C_5H_{10}O_2$. The compound indicates the peaks at 1739 cm^{-1} in the IR spectrum and the signals at δ 4.5 (multiplet, 1H), 2.0 (singlet, 3H) and 1.2 ppm (doublet, 6H) in the NMR spectrum.

25. Determine the structure of the unknown compound with the formula $C_5H_{10}O$. The IR spectrum, indicate the peak at 1705, 3000 cm^{-1}, and the signals in the NMR spectrum appears at δ 2.5 (triplet, 2H), 2.1 (singlet, 3H), 1.6 (multiplet, 2H) and 0.9 ppm (singlet, 3H).

26. A compound with molecular formula $C_8H_{11}N$ shows the following spectra, ^1H NMR (δ, ppm): 7.2 (m, 5H), 1.2 (t, 3H), 3.1 (q, 2H), 3.5 (s, 1H), IR (v_{max} cm^{-1}): 3400, 3000, 1600. Predict the structure of compound. Explain each peak of both IR and NMR.

27. An organic compound $C_{10}H_{12}O_2$ gave the following spectral data, UV (λ max): 220 nm, IR significant absorption at 3075, 2975, 1745, 1605, 1500 and 1450 cm^{-1}; ^1H NMR 2.02 (s), 2.93 (t, $J = 7$ Hz), 4.30 (t, $J = 7$ Hz) and 7.30 (s) in the intensity ratio 3 : 2 : 2 : 5. Deduce the structure of the compound explain the spectral data.

28. Account for the fact that the δ value of aromatic hydrogen ($6 – 8.5$ ppm) is higher than vinylic hydrogen (δ $4.6 – 5.9$ ppm).

29. Suggest the structure for a compound C_9H_{12} showing NMR peaks at δ 7.1, 2.2, 1.5 and 0.9 ppm.

30. An organic compound A is obtained by reaction between benzene and halohydrocarbon B in the presence of anhydrous $AlCl_3$. The ^1H NMR spectrum of A shows a triplet at δ 1.25 (3H), a quartet at 2.6 (2H), singlet at 7.2 ppm (5H). What are A, B?

31. Determine the number of signals shown by each set of compounds in ^1H NMR spectrum:

    (i) $CH_3CH_2CH_2Br$ $CH_3CH_2OCH_3$

                          CH_3OCH_3

                          CH_3OCHCl_2

$$\underset{\underset{CH_3}{|}}{\overset{\overset{CH_3}{|}}{CH_3COCH_3}}$$

$$\underset{\underset{OH}{|}}{CH_3CHCH_2CH_3}$$

$$\underset{\underset{Br}{|}}{CH_3CHCH_3}$$

(ii)

(iii)

32. How many groups of non-equivalent protons are present in the compound $CH_2=CH–CH_2OH$ and what is the ratio of the number of protons in each groups?

33. The NMR spectrum of an organic compound with molecular formula $C_3H_3Cl_5$ shows the following signals: $\delta = 4.82$ ppm, triplet (1H), $\delta = 6.07$ ppm, doublet (2H). Assign the structure.

34. Assign the structure to an organic compound with molecular formula $C_{10}H_{14}$. The ^1H NMR spectrum exhibits the following signals: $\delta = 1.30$ singlet (9H), $\delta = 7.28$ ppm, singlet (5H).

35. Assign the structure to an organic compound with molecular formula $C_{10}H_{13}Cl$. The ^1H NMR spectrum exhibits the following signals: $\delta = 1.57$ singlet (9 H), $\delta = 3.17$ singlet (2H), $\delta = 7.27$ ppm, singlet (5H).

36. Assign the structure to an organic compound with molecular formula $C_9H_{11}Br$. The ^1H NMR spectrum exhibits the following signals: $\delta = 2.15$ quintet (2H), $\delta = 2.75$ triplet (2H), $\delta = 3.38$, triplet (2H), $\delta = 7.25$ ppm singlet (5H).

37. The ^1H NMR spectrum of cyclohexane shows only a single resonance peak. As the temperature of the sample is lowered, the sharp signals peak broadens until -66.7 °C it begins to split into broad peaks. On further lowering the temperature to -100 °C, each of the two broad bands begins to give a splitting pattern of its own. Explain the origin of the peaks in this case.

38. Explain, why aromatic protons absorb at lower field than acetylenic protons.

39. Explain, why the methyl peak splits into a triplet and the methylene peak gets split into a quartet in acidified ethanol.

40. Explain, why the high resolution NMR spectrum of ordinary ethanol shows only a singlet for a hydroxyl proton while the spectrum of highly pure ethanol sample shows the multiplet for hydroxyl proton.

41. What is spin-spin coupling. What are its units? Is the coupling constant independent of the applied field or dependent on it?

42. Give a brief account on significance of the chemical shifts in the ^1H NMR studies.

43. Explain the shielding proton in the following order: ethylene > acetylene > benzene.

44. Assign the structure to the organic compound with molecular formula C_4H_9Cl with $\delta = 1.0$ doublet (6H), 2.0 multiplet (1H), 3.35 ppm doublet (2H).

45. How can the aldehydic protons be distinguished from ethylene in the ^1H NMR spectrum?

46. How the following isomeric compounds with molecular formula can be distinguished on the basis of ^1H NMR spectroscopy:

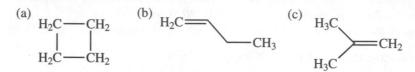

47. When methallyl chloride ($CH_2=C(CH_3)CH_2Cl$) was treated with sodamide in tetrahydrofuran solution a hydrocarbon C_4H_6 was obtained and gives the following ^1H NMR spectrum, a doublet $\delta = 0.83$, 2H, $J = 2Hz$, a doublet $\delta = 2.13$, 3H, $J = 1$ Hz, multiplet $\delta = 6.40$ ppm, 1H. Assign the structure to the hydrocarbon.

48. How many signals would you expect for the following compounds?
    (a) Allyl chloride   (b) $HC\equiv C-CH_2Br$
    (c) Allyl bromide   (d) $C_6H_5CH_2COOCH_3$

49. What are equivalent and non-equivalent protons?

50. What is chemical shift? What are factors influencing chemical shift?

51. Explain shielding and deshielding of protons?

52. What is coupling constant? What are the factors that govern the coupling constant?

53. Why is tetramethylsilane used as standard reference in NMR?

54. Assign the structure to the compound having general formula C_4H_9Br that shows the signals in the ^1H NMR spectrum at δ 1.04 (d, 6H), 1.95 (m, 1H), 3.33 (d, 2H).

55. An organic compound $C_{14}H_{22}$ shows two singlet at δ 8.2 (4H), 1.4 (18H). Assign the structure.

56. How will you distinguish between the following compounds on the basis of ^1H NMR:
    (a) 1,4-Dichlorobenzene and 1,2-dichlorobenzene.
    (b) Propionaldehyde and acetone.

57. Assign the structure and chemical shift value from the following ^1H NMR spectra with molecular formula, C_5H_{10}.

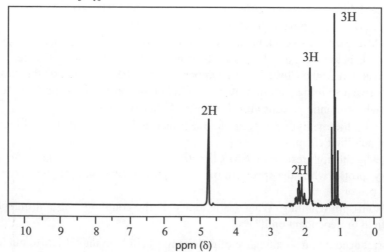

[Hint: 2-Methylbut-1-ene]

58. Assign the structure and chemical shift value from the following ^1H NMR spectra with molecular formula, $C_3H_7NO_2$.

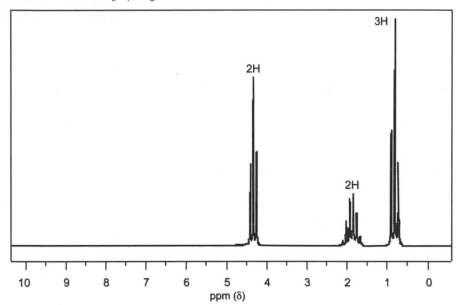

[Hint: Nitropropane]

59. The following compound is a monosubstituted aromatic hydrocarbon with molecular formula, C_8H_{10}. Assign the structure and chemical shift values.

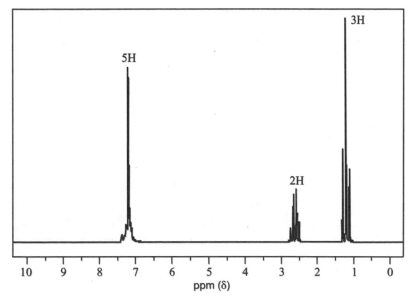

[Hint: Ethyl benzene]

60. Assign the structure and chemical shift values to following compound with molecular formula, C_2H_5Br.

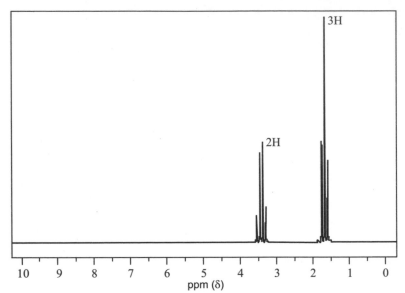

[**Hint:** Ethyl bromide]

61. The following compound is a disubstituted aromatic hydrocarbon with molecular formula, $C_{10}H_{14}$. Assign the structure and chemical shift values.

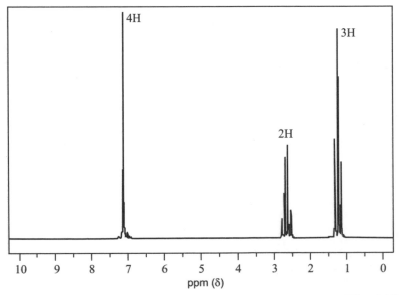

[**Hint:** 1,4-Diethyl benzene]

62. Assign the structure and chemical shift values to the bromo compound with molecular formula, $C_5H_{11}Br$ using the information available in following spectra.

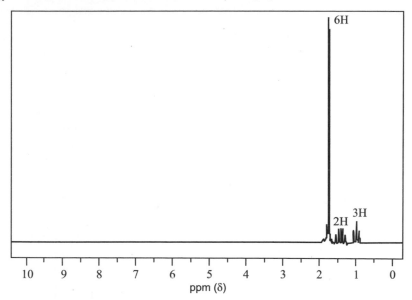

[**Hint:** 2-Bromo-2-methylbutane]

63. Assign the structure and chemical shift values to a ketone with molecular formula, $C_7H_{14}O$ using the information available in following spectra.

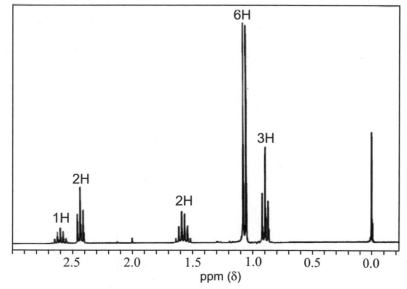

[**Hint:** 2-Methyl-3-hexanone]

64. Assign the structure and chemical shift values to an aldehyde with molecular formula, $C_7H_5NO_2$ using the information available in following spectra.

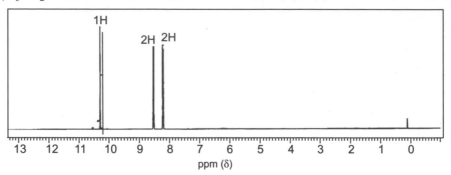

[**Hint:** 4-Nitro benzaldeyde]

65. Assign the structure and chemical shift values to the following compound with molecular formula, C_3H_8O.

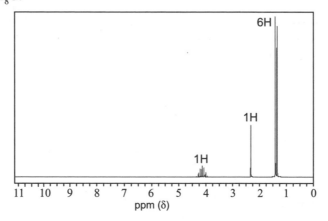

[**Hint:** 2-Propanol]

66. Assign the structure and chemical shift values to an aldehyde with molecular formula, $C_8H_8O_2$ using the following spectra.

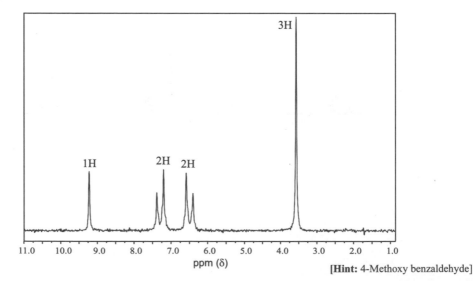

[**Hint:** 4-Methoxy benzaldehyde]

67. Assign the structure and chemical shift values to the halo-compound with molecular formula, C_4H_9Br using in following spectra.

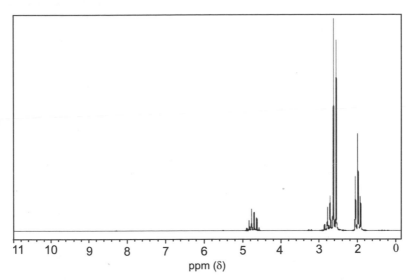

ppm (δ)

[**Hint:** 2-Bromobutane]

68. Assign the structure and chemical shift values to the halo- compound with molecular formula, $C_5H_{11}Cl$ using in following spectra.

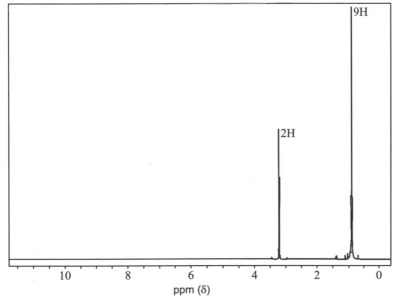

ppm (δ)

[**Hint:** 1-Chloro-2,2-dimethylpropane]

69. The following compound is a disubstituted aromatic hydrocarbon with molecular formula, $C_{10}H_{14}$. Assign the structure and chemical shift values.

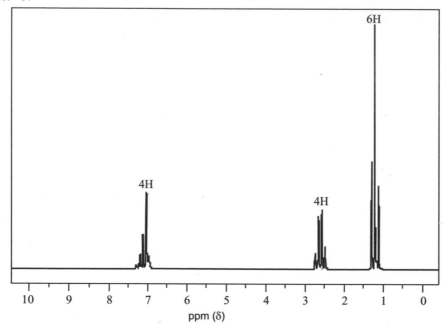

[**Hint:** *m*-Diethyl benzene]

70. Assign the structure and chemical shift values to an aldehyde with molecular formula, C_7H_6O using the following spectra.

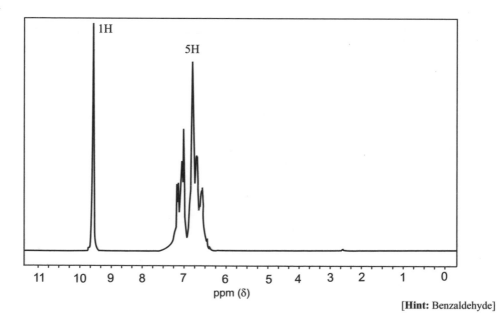

[**Hint:** Benzaldehyde]

71. Assign the structure and chemical shift values to a carboxylic acid with molecular formula, $C_4H_8O_2$ using the following spectra.

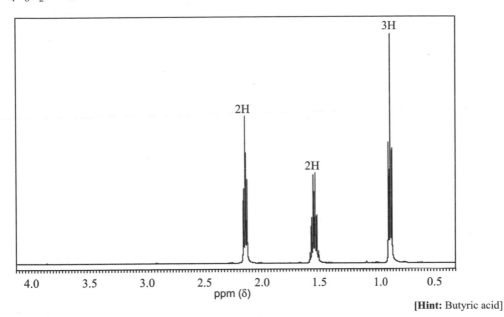

[**Hint:** Butyric acid]

72. The following compound with molecular formula, C_7H_8O is an aromatic ether. Assign it the structure and chemical shift values.

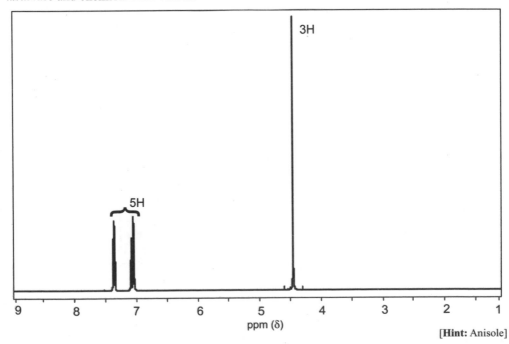

[**Hint:** Anisole]

73. The following compound with molecular formula, $C_7H_6N_2O_3$ is an aromatic amide. Assign it the structure and chemical shift values.

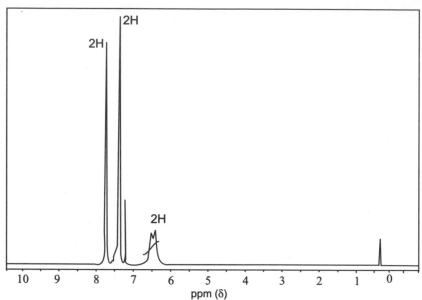

[Hint: *p*-Nitroacetanilide]

74. The following compound with molecular formula, $C_5H_{10}O_2$ is a carboxylic acid. Assign it the structure and chemical shift values.

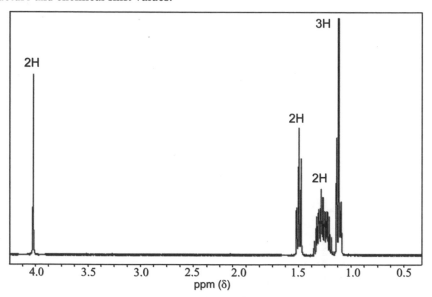

[Hint: Pentanoic acid]

Chapter

5

¹³C NMR Spectroscopy

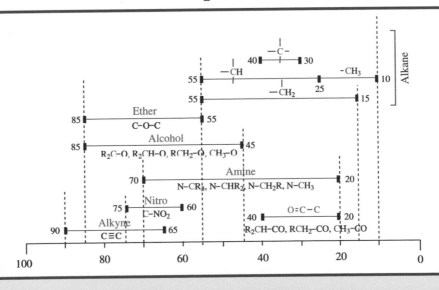

In 1957, Paul Lauterbur recorded the first spectrum at natural abundance (1.1%). ¹³C NMR provides direct information about the carbon skeleton of a molecule. The technique can be used to determine the number of non-equivalent carbon atoms and to identify different types of carbon atoms (primary, secondary, carbonyl, aromatic etc.) present in the compound.

5

^{13}C NMR Spectroscopy

5.1 INTRODUCTION

Carbon-13 nuclear magnetic resonance (NMR) spectroscopy is the study of carbon nuclei through NMR, which determines the structure of the organic compounds. This along with proton nuclear magnetic resonance (^{1}H NMR) and infrared (IR) spectroscopy gives the complete structural information of an unknown compound. The technique can be used to determine the number of non-equivalent carbon atoms and to identify different types of carbon atoms (primary, secondary, carbonyl, aromatic etc.) present in the compound. Thus, ^{13}C NMR provides direct information about the carbon skeleton of a molecule.

5.2 NATURAL ABUNDANCE OF CARBON-13

The natural abundance of Carbon-12 is 98.9% but its nuclear spin (I) is zero and therefore is NMR inactive, whereas, the natural abundance of ^{13}C is 1.1%, and has I = ½, and is NMR active. The resonance of ^{13}C nuclei are more difficult to observe than those of protons (^{1}H) as only 1.1% of the carbon atoms in a molecule are detectable by ^{13}C NMR, although this can be overcome by isotopic enrichment. The resonance frequencies for ^{13}C are about ¼ those of ^{1}H (the ratio of the γ values for these nuclides). Thus, for 7.05 T magnet the ^{13}C resonance frequency is about 75.5 MHz, while for proton it is 300 MHz (**Table 1**). Also in ^{13}C NMR, carbon and hydrogen (typically from 100 to 250 Hz) undergoes a large one bond J-coupling constants due to which the spectra is highly complicated and in order to suppress these couplings, carbon NMR spectra are proton decoupled to remove the signal splitting. Due to large range and the sharpness of decoupled peaks, impurities are readily detected and thus the mixture may be readily analysed. Even stereoisomers that are difficult to analyse by means of ^{1}H spectroscopy usually show discrete ^{13}C peaks. Coupling between carbons can be ignored due to the low natural abundance of ^{13}C. Hence, in contrast to ^{1}H NMR spectra which show multiplets for each proton position, carbon NMR spectra show a single peak for each chemically non-equivalent carbon atom. Also in ^{1}H NMR, the intensities of the signals are not normally proportional to the number of equivalent ^{13}C atoms and are instead strongly dependent on the number of surrounding spins (typically ^{1}H). Spectra can be made more quantitative if necessary by allowing sufficient time for the nuclei to relax between repeat scans. ^{13}C chemical shifts follow the same principles as those of ^{1}H, although the typical range of chemical shifts is much larger than for ^{1}H (by a factor of about 20).

Table 1. Comparison of ^{1}H and ^{13}C spectrometer frequencies (in MHz).

^{1}H frequency (MHz)	Magnetic field (B_O)/tesla	^{13}C frequency (MHz)
100	2.3	25.14
200	4.7	50.28
300	7.1	75.4
500	11.7	125.7
600	14.1	150.8

5.3 SOME USEFUL TERMS

5.3.1 Referencing ^{13}C NMR Spectra

Tetramethylsilane (TMS) is the primary reference for ^{13}C spectra. The relatively low sensitivity of ^{13}C NMR requires the addition of substantial amounts of TMS, so it is common to use solvent peaks as a secondary reference. Chemical shifts of several common solvents used in NMR spectroscopy are given below (**Table 2**).

Table 2. Chemical shifts of several common solvents used in NMR spectroscopy.

S. No.	Solvent	^{13}C NMR Chemical Shift
1.	Acetic Acid	179.0 (1), 20.0 (7)
2.	Acetone	206.7 (13), 29.9 (7)
3.	Acetonitrile	118.7 (1), 1.39 (7)
4.	Benzene	128.4 (3)
5.	Chloroform	77.2 (3)
6.	Dimethyl Sulfoxide	39.5 (7)
7.	Methanol	49.1 (7)
8.	Methylene Chloride	54.0 (5)
9.	Pyridine	150.3 (1), 135.9 (3), 123.9 (5)
10.	Water (D$_2$O)	–

5.3.2 Chemical Shift Equivalence

Chemical shift equivalence for protons also applies to the carbon atoms and the chemical equivalence or non-equivalence of carbon atoms is judged in the same way as for protons. For example, all the three carbons in *t*-butyl acetate are equivalent and thus the spectra consists of five different signals (**Figure 1**).

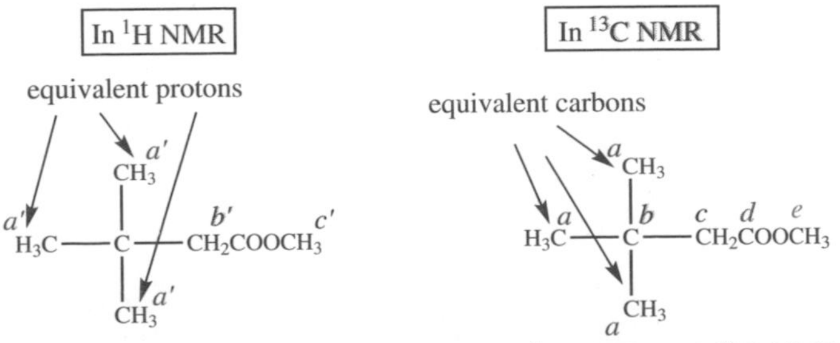

Figure 1. Comparison of number of peaks observed in ^{1}H NMR and ^{13}C NMR spectra of *t*-butyl acetate.

5.3.3 Chemical Shifts

Chemical shifts in ^{13}C NMR spectroscopy ranges to about δ 0 to 250 ppm with respect to TMS (commonly used reference compound). On moving downfield from TMS, the sequence alkane, substituted alkanes, alkynes, olefins, aromatic and aldehydes for ^{13}C chemical shifts are similar to that as in case of ^1H NMR spectroscopy (**Scheme 1** and **2**).

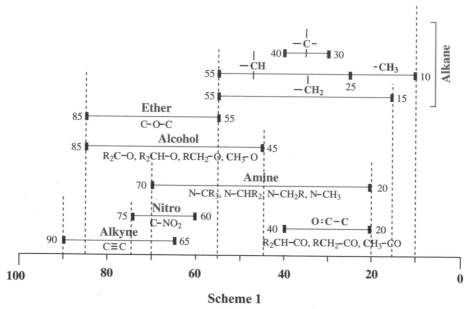

Scheme 1

The order of ^{13}C chemical shift for various types of compound is:

C=O (Aldehyde and ketone) > C=O (carboxylic acid, esters and amides) > alkene, nitrile and aromatic compound > alkynes > C–O (alcohol and ether) > C–X (X = Cl, Br, N) > alkanes (**Scheme 1** and **2, Table 3**).

Table 3. Chemical shifts (δ) of different functional groups.

Type of Carbon		Chemical shift (δ in ppm)
Alkanes	$R-CH_3$	10–25
	R_2-CH_2	15–55
	R_3-CH	25–55
	R_4-C	30–40
Alkene	$R_2C=CR_2$	100–170
Alkyne	$RC\equiv CR$	65–90
Aromatic	⬡	120–150
Ketone	R_2-CO	190–220
	Ar–CO–R Ar–CO–Ar =CO–R	

Type of Carbon		Chemical shift (δ in ppm)
Aldehyde	—C=O | H	180–220
Nitrile	C≡N	110–125
Alcohol	R–OH	45–85
Carboxyl $\begin{bmatrix} O \\ \parallel \\ R^{C}X \end{bmatrix}$	X = OH	165–185
	X = OR′	160–175
	X = NH₂ or NHR′	150–175
	X = Cl	165–175
Amine	R_3C–N R_2CH–N RCH_2–N	20–70

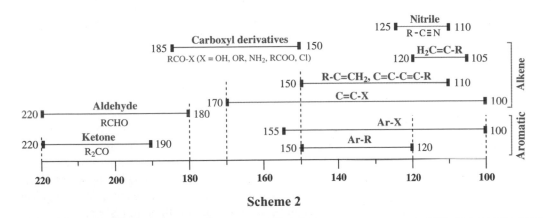

Scheme 2

EXERCISE PROBLEMS 5.1

1. What is the basic principle of Carbon-13 nuclear magnetic resonance? What information ¹³C NMR provides about the structure of the molecule.

2. What is the natural abundance of carbon-13?

5.4 FACTORS AFFECTING ¹³C CHEMICAL SHIFTS

5.4.1 Inductive Effect

Electronegative groups (α-effect of O, N, F, Cl, Br) cause large ¹³C chemical shift effects, which can be used to classify groups of resonances (**Table 4**). Within groups there are smaller effects which are useful, for example, resonance interactions within the systems cause predictable upfield and downfield chemical shift effects. Heavy atoms (e.g., iodine, tellurium) cause upfield shifts, as does the accumulation of adjacent sterically crowded carbon atoms (branching effects).

Table 4. Influence of functional group X on the chemical shift position of nearby carbons in alkene group and benzene rings.

Functional group		CH_3	$CH=CH_2$	$C\equiv CH$	C_6H_5, Ar	F	Cl	Br	I	NH_2	NHR	NR_2	NO_2	NHCOR, NRCOR	CN	SH	OH	OR	OCOR	COOH, COOR, $CONR_2$	COR, CHO	SO_3H, SO_2NR_2
X\ $C_1=C_2$ Alkene (Base value $=\delta\ 123$)	C–1	10	15	–	13	25	3	–8	–38	–	–	–	22	–	–15	–	–	29	18	4	14	–
	C–2	–8	–6	–	–11	–34	–6	–1	7	–	–	–	–1	–	15	–	–	–39	–27	9	13	–
X ipso C o m p Benzene (Base value $=\delta\ 128$)	ipso	9	9	–6	13	35	6	–5	–32	18	20	22	20	10	–16	4	27	30	23	2	9	16
	o	0	0	4	–1	–14	0	3	10	–13	–14	–16	–5	–7	4	1	–13	–15	–6	2	1	0
	m	0	0	0	1	1	1	1	2	3	1	1	1	1	1	1	1	1	1	0	1	0
	p	–2	–2	0	–1	–5	–2	–2	–1	–10	–10	–10	6	–4	6	–3	–7	–8	–2	5	6	4

5.4.2 Effect of Hybridisation

Hybridisation (sp^3, sp^2, sp) of the carbon has a significant effect on the chemical shift. The sp^3 hybridised carbon appears the most upfield, followed by sp and then sp^2, for example, the sp^3 hybridized alkanes absorb from δ –5 to 55 ppm, alkynes aborb near to δ 65–90 ppm and alkenes aborb near δ 110–150 ppm. However, sp^2 carbons of the double bond type cannot be distinguished from those of the aromatic type. This is in contrast to the situation with proton NMR.

5.4.3 Effect of Substituent on Alkanes

Alkanes absorb from –5 to 55 ppm that depend on the presence of the substituent. ^{13}C chemical shifts are significantly affected by the presence and number of substituents at α, β and γ-positions with respect to the carbon under study (**Table 5**). The effects are largest for substituent changes at the carbon itself (α-effect) but sizable substituent effects are seen at the β, γ and δ position. Generally, the α-effect are strongly dependent on electronegativity of the substituent, the β-effect shifts the absorption all to higher frequency, and fairly of similar in size, and γ-effect shifts the absorption all to lower frequency (except for organometallic substituents).

5.4.3.1 α-Substituent effect

If the hydrogen attached to the carbon under study is substituted by some group, than a change in its resonance takes place depending on the nature of the group attached, this is known as α-effect. Thus, there is an increase in chemical shift (δ value) on moving from primary, secondary, tertiary and quaternary carbon atoms with an average increase per hydrogen-atom replaced in the range of δ 7–10 ppm. For example, CH_4 resonates at δ 2.6 ppm, CH_3–CH_3 at 5.7 ppm, $(CH_3)_2CHCH_3$ at 9.1 ppm and $(CH_3)_3CCH_3$ at 28.1 ppm. Thus, from the above values it can be concluded that the increase in alkyl substituents on the carbon under study increases its δ value. As compared to the α-methyl group attached to the carbon under study the presence of polar group has much larger downfield shift because of the inductive effect.

$$CH_3-CH_2-CH_2-CH_2-H$$
$$CH_3-CH_2-CH_2-CH_2-Cl$$

For example, the chlorine causes the α-carbon a large downfield shift from 13.67 to 44.3 ppm, a difference of +30.6 ppm. Similarly the α-effect exerted by various substituents leading to a downfield shift (**Table 5**).

Table 5. The α-effect exerted by various substituents leading to a downfield shift.

F	Br	Cl	NH₂	OH	NO₂
70.1	19.3	30.6	29.7	48.3	64.5

5.4.3.2 β-Substituent effect

If, the substitution is attached at β-position of the carbon under study, it results in the larger effect on the δ value. Almost all substituents cause substantial high-frequency shifts in the δ value (usually ~9 ppm, smaller if crowded), and these are not very dependent on the electronegativity of the perturbing substituent. Chlorine causes the β-carbon a large downfield shift from 22.6 to 32.7 ppm, a difference of 10.1 ppm. Also, in case of propane the δ value is 16.3 ppm whereas in *iso*-butane it is 25.4 ppm, there is a downfield shift of 9.1 ppm due to the replacement of hydrogen on an adjacent carbon by methyl group. This is β-effect, and it has a value of +9.5 ppm.

5.4.3.3 γ-Substituent effect

The replacement of an H by X on the second carbon atom from the carbon under study leads to γ-effect. If, the γ-carbon has attached hydrogen, then only all X-substituents show γ-effect. The γ-effect is extensively used for stereochemical assignments. If a γ-atom is close to a carbon in one isomer, and remote in another, then that carbon will be upfield in the first isomer.

Table 6. Effects of conjugation on the chemical shift of the carbonyl carbon.

Carbonyl compound	δ Values C=O (ppm)	Conjugated carbonyl compound	δ Values C=O (ppm)
	206		198
	202		194
	174		166.0
	220		210

Carbonyl compound	δ Values C=O (ppm)	Conjugated carbonyl compound	δ Values C=O (ppm)
(cyclohexanone)	210	(cyclohexenone)	199
(NC—propyl)	121	(NC—vinyl/allyl)	118

5.4.4 Effect of Substituent on the Chemical Shift of Carbonyl Carbons

Carbonyl groups of ketones and aldehydes appear in the range δ 190–220 ppm, while esters, acids, amides and related carbonyl functions appear in between δ 150 and 175 ppm (**Table 6**). The ketone region is quite distinctive, whereas, the various carboxylic acid derivatives do not have distinct chemical shift ranges. Thus, acids, esters, acid chlorides, amides, anhydrides are not readily distinguished in the ^{13}C NMR spectrum. Also, carbonates, ureas, and carbamates are not well separated from the carboxylic acid derivatives.

5.4.4.1 Effect of conjugation

In case of carbonyl compound the conjugation to a double bond or aromatic ring leads to upfield shift of δ 6–10 ppm (**Table 6**). It may be attributed to the charge delocalisation by the aromatic ring or double bond makes the carbonyl compound less electron deficient. Similar effect is observed in nitriles but with a smaller magnitude.

5.4.4.2 Effect of hydrogen bonding

Intramolecular hydrogen bonding leads to downfield (larger δ value) shift. Most carbon signals are insensitive to solvent effect, but carbonyl groups move downfield in protic solvents, which can be attributed to hydrogen bonding (**Table 7**).

Table 7. Effect of hydrogen bonding on the chemical shift of the carbonyl carbon.

Carbonyl compound	δ Values C=O (ppm)	H-bonding in carbonyl compound	δ Values C=O (ppm)
(benzaldehyde)	192	(salicylaldehyde)	198
(acetone)	206	(acetol/hydroxy)	212
(methyl ethyl ketone)	206	(hydroxy ketone)	210

5.5 CALCULATION OF CHEMICAL SHIFT IN ALKANES

The substituent on the carbon shifts the signal more downfield as compared to the corresponding shift in the proton spectra. Analysis of the ^{13}C chemical shift of acyclic alkanes led to the first accurate method for the prediction of chemical shift. Grant-Chaney calculations are based on the observation that, in addition to α, β, γ and δ effects, there are predictable branching effects, such that the α and β-effects which are nearly constant for linear molecules, become progressively smaller when there are nearby tertiary and quaternary carbons (**Table 8**).

Table 8. Influence of functional group on the chemical shift of nearby carbon in alkane chain.

Substituents attached to carbon $X-C-C-C-C$ (α β γ)		CH_3	Cyclohexane Axial CH_3	Cyclohexane Equatorial CH_3	$CH=CH_2$	$C\equiv CH$	C_6H_5, Ar	F	Cl	Br	I	NH_2, NHR, NR_2	NO_2	NH_3^+	CN	SH	OH	OR	OCOR	COOH, COOR	COR, CHO
α	1°	9	1	6	22	4	23	70	31	19	−7 to 20	29	62	25	3	2	50	50	52	20	30
	2°	6	–	–	16	–	17	–	35	28	–	24	–	–	4	–	45	24	50	16	24
	3°	3	–	–	12	–	11	–	42	37	–	18	–	–	–	–	40	17	45	13	17
β		9	5	9	7	3	10	8	10	11	11	11	3	7	2	2	9	10	7	2	2
γ		−3	−6	0	−2	−3	−3	−7	−5	−4	−2	−4	−5	−3	−3	−2	−3	−6	−6	−3	−3

The α and β effects lead to a deshielding effect, whereas, γ-effect leads to a shielding effect of −2.5 ppm, δ effect has +0.3 ppm and ε-effect is +0.1 ppm. Thus, the chemical shift for carbon atom in alkanes can be calculated using the following equation:

$$^{TMS}\delta_c = -2.5 + 9.1\ n\alpha + 9.4\ n\beta - 2.5\ n\gamma + 0.3\ n\delta$$

Where, $^{TMS}\delta_c$ = predicted chemical shift for a carbon with reference to TMS

$n\alpha$, $n\beta$, $n\gamma$, $n\delta$ = number of carbon atoms at one, two, three and four bonds from carbon atom whose chemical study is being calculated.

The δ value for ^{13}C can also be calculated using the equation given below:

$$^{TMS}\delta_c = 1 + 7n_1 + 8n_2 - 2n_3$$

Where, $^{TMS}\delta_c$ = predicted chemical shift for a carbon with reference to TMS

n_1, n_2, n_3 = number of carbon atoms at one, two and three bonds away from the carbon atom under study.

EXERCISE PROBLEMS 5.2

1. What do you mean by chemical shift equivalence?
2. What is the range of chemical shifts in ^{13}C NMR spectroscopy?
3. What are the factors which affect the ^{13}C NMR spectroscopy?

5.6 MODE OF RECORDING ^{13}C NMR SPECTRA

5.6.1 Proton Noise Decoupling

Generally the ^{13}C NMR spectra are obtained as proton decoupled spectra. In the decoupled ^{13}C spectra all the interactions between ^{13}C and protons are nullified by the decoupling techniques and therefore singlet for each chemically non-equivalent carbon atom present in the molecule is observed. It provides information to count the number of different types of carbons and predicts their environment for chemical shifts. Due to the decoupling the entire spectrum gets simplified but information on attached hydrogen is lost.

Proton decoupling is carried out by simultaneously irradiating all the protons in the molecule with a broad spectrum of frequencies. On irradiation protons become saturated and they undergo rapid upward and downward transitions among all possible spin states, due to which decoupling of any spin-spin interactions between ^{13}C and hydrogen takes place. By these rapid changes all spin interactions are averaged to zero. Thus, the carbon nucleus senses only one average spin state for the attached hydrogens rather than two or more distinct spin states.

5.6.2 Off Resonance Decoupling

In an off resonance decoupled ^{13}C spectrum, the coupling between carbon atom and directly attached hydrogen atom is observed. A carbon bonded to n protons gives a signal that splits into $n +1$ peak, indicating the number of protons attached to each kind of carbon. It reduces the magnitude of coupling constant and retains the coupling between the carbon and directly attached proton (^{1}H) but effectively removes the coupling between the carbon atom and the more remote protons. As a result the multiplet become narrow and the spectrum now appears as a series of singlets, doublets, triplets and quartet depending on whether the carbon is quaternary, or has one, two or three protons attached. Vicinal and long range couplings collapse but the geminal carbon and hydrogen couplings remain.

In off resonance (double resonance) the second radio frequency transmitter is set either upfield or downfield from the usual sweep width of a normal proton spectrum (off resonance). In contrast frequency of the decoupler is set to coincide exactly with the range of proton resonance in the decoupling experiment. Also, the power of the decoupling oscillator is held low to avoid complete decoupling.

5.6.3 Distortionless Enhancement by Polarisation Transfer (DEPT) ^{13}C Spectra

DEPT is a very useful method for determining the presence of primary, secondary and tertiary carbon atoms. The DEPT experiment differentiates between CH, CH_2 and CH_3 groups by variation of the selection angle parameter. If the angle is 45°, it gives all carbons with attached protons (regardless of number) in phase and the spectra are known as DEPT-45 spectra. Secondly, if the angle is 90° only a carbon bonded to one hydrogen atom i.e., CH groups will appear in the spectrum and the others will be suppressed and this spectrum is known as DEPT-90 spectra. Thirdly, if the angle is 135° all carbons attached provide a signal, but the phase of the signal will be different, depending on the number attached hydrogen is an odd or even number such spectra is known as DEPT-135 spectra. Signals arising from odd number of hydrogens attached, i.e., CH and CH_3 will appear as positive absorption in the spectrum while carbon attached to even hydrogens (CH_2) appear as negative absorptions and no peak will be observed for the quaternary carbon. Signals from quaternary carbons and other carbons with no attached protons are always absent (due to the lack of attached protons).

It is important to understand that the appearance of positive and negative signals can be reversed by phasing, so it is necessary to have some way of determining whether the spectrum has been phased for CH_2 positive or negative. "Leakage" can occur in DEPT-90 spectra because ^1JC–H varies as a function of environment, and the technique assumes that all ^1JC–H are identical. This can result in small peaks for CH_2 and CH_3 signals, which should have zero intensity. For similar reasons the C–H of terminal acetylenes (C≡C–H) will show anomalous intensities in DEPT spectra (either nulled or very small in DEPT-90, or present in DEPT-135) because the CH coupling is much larger (around 250 Hz) than the normal value of 125 Hz for which the DEPT experiment is usually parameterised.

5.6.4 Gated Decoupling

In this mode area under each peak is directly proportional to the number of carbon atoms causing that peak. By electronically integrating the area under each peak, the relative number of carbon atoms represented by each peak can be determined in the same way as in PMR spectroscopy. Gated decoupled spectra contain more information than normal proton noise decoupled spectra because the former can be integrated. As the NMR spectrometer is two or three times less sensitive in the gated-decoupling mode than in the proton noise-decoupling mode. Therefore, more time and large amount of the sample is required to obtain the spectrum. Thus, gated coupled spectra are rarely used. Some carbon atoms give rise to exceptionally small signals due to weak NOE enhancement which is one of the reasons for wrong relative intensities of the signals. In gated-decoupling mode, the noise decoupler is gated (switched) on during the pulse at the early part of free induction decay (FID), and then gated off during the pulse delay. Thus, NOE enhancement is minimised for all carbon atoms. This happens because the free induction signal decays quickly in an exponential manner, whereas the NOE factor, longer signal acquisition time is required which is a demerit of the gated decoupling mode.

5.6.5 Spin-spin Coupling

Spin-spin coupling is less important in CMR spectra than PMR spectra, as spectra are usually noise decoupled. However, in off-resonance decoupling mode, coupling due to protons directly attached to ^{13}C atoms (^{13}C–H) are observed but other coupling (^{13}C–C–H, ^{13}C–C–C–H) are removed. ^{13}C–H coupling constants ($^1J_{CH}$) range from about 110–320 Hz. Coupling of ^{13}C to ^1H has also been observed over two (^{13}C–C–H) or three (^{13}C–C–C–H) bonds, J values denoted as $^2J_{CH}$ and $^3J_{CH}$ respectively. The magnitude of coupling constants increases with increased s-character of ^{13}C–H bond. $^2J_{CH}$ values usually ranges from about –5 to 60 Hz, whereas, $^3J_{CH}$ ranges from –5 to 25 Hz. However, in aromatic rings, $^3J_{CH}$ values are characteristically larger than $^2J_{CH}$ values in benzene, $^3J_{CH} = 7.4$ Hz, whereas, $^2J_{CH} = 1.0$ Hz. Coupling of ^{13}C to other nuclei, e.g., D, ^{19}F etc. may be observed in proton noise decoupled spectra. The ^{13}C–^{13}C coupling is usually not observed because of the low probability of two adjacent ^{13}C atoms in a molecule.

═══════ EXERCISE PROBLEMS 5.3 ═══════

1. What is proton noise decoupling?

2. What do you mean by the term off resonance decoupling?

3. What is (DEPT) ^{13}C spectra?

5.7 APPLICATIONS OF ^{13}C NMR SPECTROSCOPY

5.7.1 Detection of a Double Bond and Distinction between *cis* and *trans* Alkenes

The γ-effect can be used for distinguishing E and Z isomers of alkenes, especially trisubstituted ones, where other techniques (such as ^3JHH) are not available. The carbon atoms directly attached to *cis*-isomer are more shielded than those with the *trans*-isomer. A methyl (or other carbon) group which is *cis* to a substituent will appear upfield of the isomer in which the carbon substituent is *cis* to a hydrogen.

the carbon shift will be upfield

X = C, O, N, S, halogen

The γ-effect

5.7.2 Determination of Acyclic *syn-anti* Stereochemistry

The γ-interactions present in axial substituents provide the basis for configurational assignment of *syn* and *anti* 1,3-diols using the methyl group chemical shifts of their acetonide derivatives. In the *syn* isomers of the acetonides the 6-membered ring has a well-defined chair conformation, with both R-substituents equatorial. This places one of the acetonide methyl groups axial, the other equatorial, leading to a 10 ppm shift difference between the two methyls. The *anti* acetonides have a twist boat conformation, which places the two methyls in a very similar environment, and hence there is a very small chemical shift difference between them.

In addition to providing configurational information for systems with well-defined *gauche/anti* or *cis/trans* relationships, the generalised upfield shifts of all four of the atoms involved in *gauche* interactions can also be the basis for stereochemical assignments of diastereomeric pairs in acyclic systems. Thus *syn* and *anti* 1,3-diols show a well defined upfield shift for C–O carbons in the *anti* compared to the *syn* isomer. The rationale for this behavior is that intramolecularly H-bonded conformations place a substituent in a pseudo-axial orientation in the *anti* isomers, hence upfield shifts, whereas all substituents can be equatorial in the *syn* isomer.

SOLVED PROBLEMS

Q1. Although the natural abundance of Carbon-12 is 98.9% but still it is NMR inactive. Explain.

Sol. The natural abundance of Carbon-12 is 98.9% but its nuclear spin (I) is zero and therefore is NMR inactive.

Q2. Why for 7.05 T magnet the ^{13}C resonance frequency is about 75.5 MHz while for proton it is 300 MHz?

Sol. The gyromagnetic ratio of ^{13}C is only 1/4 that of ^1H, which reduces the sensitivity. The resonance frequencies for ^{13}C are about 1/4 those of ^1H (the ratio of the γ values for these nuclides). Thus for 7.05 T magnet the ^{13}C resonance frequency is about 75.5 MHz while for proton it is 300 MHz.

Q3. Why carbon NMR spectra are proton decoupled?

Sol. In ^{13}C NMR carbon and hydrogen undergoes a large one bond J-coupling constants due to which the spectra is highly complicated and in order to suppress these couplings, carbon NMR spectra are proton decoupled to remove the signal splitting.

Q4. How will you distinguish between *cis* and *trans* 2-butenes on the basis of ^{13}C NMR spectroscopy?

Sol. The stereochemical influence of γ-effect in alkenes is extremely useful in assigning configuration to geometric isomers. These γ-effects are both upfield; $\delta -7.3$ ppm for *cis*-isomer and $\delta -1.9$ ppm for the *trans*, i.e., γ-effect for *cis* is stronger by δ 5.4 ppm and consequently the *cis* and *trans* 2-butenes can easily be distinguished from chemical shifts of methyl carbons, i.e., 11.4 i.e., δ 11.4 ppm in *cis* and δ 16.8 ppm in *trans*.

Q5. How will you distinguish among the carbonyl isomers pertaining to the molecular formula C_4H_8O on the basis of ^{13}C NMR spectroscopy?

Sol. The three possible isomers are:

$$CH_3COCH_2CH_3 \qquad CH_3CH_2CH_2CHO \qquad (CH_3)_2CHCHO$$
$$I \qquad\qquad\qquad II \qquad\qquad\qquad III$$

(i) The ketone (I) exhibits strongly downfield singlet, whereas, both the aldehydes (II and III) display a strongly downfield doublet.

(ii) The aldehyde II shows two triplets and one quartet, whereas, the aldehyde (III) exhibits a second doublet and a quartet.

UNSOLVED PROBLEMS

1. Which list below gives only spin active nuclei?
   (a) ^1H, ^2H, ^{12}C (b) ^2H, ^{12}C, ^{19}F
   (c) ^1H, ^{13}C, ^{19}F (d) ^1H, ^{12}C, ^{19}F

2. Which carbon environment and ^{13}C chemical shift range are *correctly* matched in the pairs below?
   (a) Nitrile C≡N; δ +50 to +100 ppm
   (b) Ketone R_2C=O; δ +190 to +230 ppm
   (c) Aliphatic group RCH_2X; δ +100 to +150 ppm
   (d) Alkene R_2C=CR_2; δ +50 to +80 ppm

3. The ^{13}C NMR spectrum of a compound A contains two signals and in the ^1H NMR spectrum there is a singlet. Which compound is consistent with these data?
   (a) Bromoethane   (b) Ethanol   (c) Dichloromethane   (d) Acetone

4. The ^{13}C NMR spectrum of a compound D shows three signals, all in the region δ = 10 to 60 ppm. Based on these data, which conclusion is incorrect?
   (a) Compound D is likely to contain aliphatic C atoms.
   (b) Compound D is not a carboxylic acid.
   (c) Compound D contains three C atoms.
   (d) Compound D does not contain a ketone functional group.

5. Describe the difference between proton-coupled (off-resonance decoupling) and proton-decoupled ^{13}C NMR spectra.

6. Describe the proton decoupled and proton coupled ^{13}C NMR spectra of 2-methylbutane.

7. How does the proton coupled signal for tetramethyl silane a standard also used for ^{13}C NMR spectroscopy appear?

8. Why is coupling between bonded ^{13}C NMR's not a factor in ^{13}C NMR spectroscopy?

9. Use the substituent effects to predict the relative values of carbon in propane (A), butane (B), isobutane (C).

10. Compare the relative order of chemical shifts of the different H's in the ^1H NMR spectrum with that of the different carbons in the ^{13}C NMR spectrum.

11. Compare the chemical shift ranges of ^{13}C NMR and ^1H NMR spectra.

12. Give the number of signals and write the structural formula for each of the following compounds:
    (i) Cyclohexene (ii) 1-Methylcyclohexene (iii) 1,3,5-Trimethylbenzene (iv) 2-Methyl-2-butene

13. Identify each of the following compounds from the given data:
    (i) C_8H_8O 71.2, t; 33.1, t; 20.3, t; 14.6, q
    (ii) $C_6H_{14}O$ 75.1, d; 35.3, s; 25.8, q; 18.2, q
    (iii) $C_6H_{14}O$ 65.5, d; 49.2, t; 25.1, d; 24.3, q; 22.7, q
    (iv) $C_5H_{11}ClO$, 62.2, t; 45.4, t; 32.9, t; 32.1, t; 23.7, t
    (v) $C_6H_{12}O_2$, 71.1, d; 37.9, t

14. Use ^{13}C NMR spectroscopy to distinguish between (i) the carbonyl isomers (ii) the alkenes arising from dehydration of $CH_3CH_2CH(OH)CH_2CH_2CH_3$, and (iii) the isomers of C_3H_8O.

15. Explain the single methyl signal in the proton-decoupled ^{13}C NMR spectrum of *cis* 1,2-dimethylcyclohexane.

16. Use the ^{13}C NMR to distinguish between the two diastereomers of 1,3,5-trimethylcyclohexane.

17. Why the strength of magnetic field required for carrying out the spectra is lower in comparison to ^1H NMR?

18. Explain why $CDCl_3$ exhibits a triplet at $\delta = 76, 77, 78$ ppm in its ^{13}C NMR spectrum.

19. How would you study hydrogen bonding in organic compounds with the help of ^{13}C NMR spectroscopy?

20. How can you distinguish between two isomeric hydrocarbons with molecular formula C_3H_4 by ^{13}C NMR spectroscopy?

21. How can you distinguish between *cis-* and *trans-*2-butenes on the basis of ^{13}C NMR spectroscopy?

22. In each of the following spectral data given suggest the structure of the compound: Multiplicity is disregarded when given the number of signals. When more than one isomer is possible, explain what other information in the spectra can be used to distinguish them.

   (i) C_2H_6O, a single ^1H and ^{13}C NMR signal

   (ii) C_3H_9N, a single ^1H and ^{13}C NMR signal

   (iii) $C_3H_6O_2$, an ester two ^1H and three ^{13}C NMR signal

   (iv) C_4H_6, (noncyclic) a single ^1H and two ^{13}C NMR signal

   (v) C_5H_{12}, a single ^1H and two ^{13}C NMR signal

   (vi) C_5H_{10}, (cyclic) one ^1H and two ^{13}C NMR signal

   (vii) C_5H_{10}, a single ^1H and ^{13}C NMR signal

   (viii) C_3H_3Cl, two ^1H and three ^{13}C NMR signal

   (ix) $C_4H_{10}O$, two each for ^1H and ^{13}C NMR signal

23. How many peaks would appear in the ^{13}C NMR spectrum of this compound? Give the approximate positions of these peaks.

24. A $C_8H_4N_2$ compound shows a sharp infrared absorption at 2230 cm^{-1}. It's ^1H NMR spectrum has a singlet at δ 7.6 ppm. The ^{13}C NMR spectrum shows three signals at δ 132, 119 and 117 ppm. Suggest a structure for this compound.

25. A $C_5H_{12}O_2$ compound has strong infrared absorption at 3300 to 3400 cm^{-1}. The ^1H NMR spectrum has three singlets at δ 0.9, 3.45 and 3.2 ppm; relative areas 3 : 2 : 1. The ^{13}C NMR spectrum shows three signals all at higher field than δ 100 ppm. Suggest a structure for this compound.

26. A $C_9H_{12}O$ compound has strong infrared absorption at 3300 to 3400 cm^{-1}. The ^{13}C NMR spectrum of this compound has six discrete signals. It's ^1H NMR spectrum has three sets of lines: Singlets at δ 1.1 (6H), 1.9 (1H) and 7.3 (5H) ppm. Suggest a structure for this compound.

27. A $C_4H_8O_2$ compound has a strong infrared absorption at 1150 cm^{-1}, but no absorption at 3300 to 3400 cm^{-1}. It's ^1H NMR spectrum shows a singlet at δ 3.55 ppm. The ^{13}C NMR spectrum shows one signal at δ 66.5 ppm. Suggest a structure for this compound.

28. A $C_{10}H_{14}$ compound in the ^1H NMR spectrum has two singlets at δ 2.45 and 7.0 ppm (ratio = 6 : 1). The ^{13}C NMR spectrum shows three signals at δ 132.9, 130.5 and 18.9 ppm. Suggest a structure for this compound.

29. A $C_{14}H_{22}$ compound exhibits ^1H NMR spectrum with two singlets at δ 1.1 and 7.25 ppm (ratio = 9 : 2). The ^{13}C NMR spectrum shows four signals at δ 147, 125, 39.3 and 30.8 ppm. Suggest a structure for this compound.

30. A compound with molecular formula $C_9H_{12}O_3$, has strong infrared absorption near $1100\ cm^{-1}$. Its ^1H NMR spectrum has sharp singlet peaks at δ 3.6 and 6.6 ppm (intensity ratio 3 : 1). Its ^{13}C NMR spectrum shows three lines at δ 165, 115 and 55 ppm. Suggest a structure for this compound.

31. A compound with molecular formula $C_9H_{18}O$, has a strong infrared absorption at $1710\ cm^{-1}$. Its ^1H NMR spectrum has a single sharp peak (a singlet) at δ 1.2 ppm. Its ^{13}C NMR spectrum shows three lines at δ 210, 45 and 25 ppm. Suggest a structure for this compound.

32. A compound has a molecular formula of $C_5H_8O_2$ and exhibits the following ^{13}C NMR spectrum: δ 199, 197, 29, 23, 6. Which of the compounds listed below would be consistent with this structure?

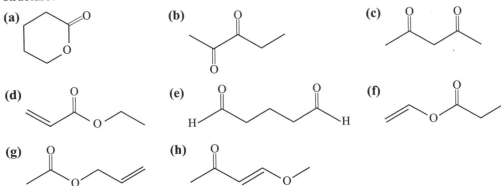

33. A compound exhibits the following ^{13}C NMR spectrum: δ 171.42, 69.47, 29.84, 22.31, 19.06. Which of the compounds listed below would be consistent with this structure?

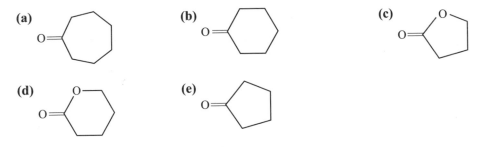

34. A compound has a molecular formula of C_8H_{10} and exhibits the following ^{13}C NMR spectrum; δ 136.42, 129.63, 125.85, 19.66. Which of the compounds listed below would be consistent with this structure?

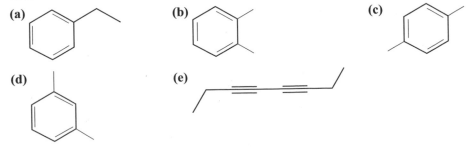

35. Correlate the different types of carbon atoms present in given organic compound by analysing the spectra given below:

(a)

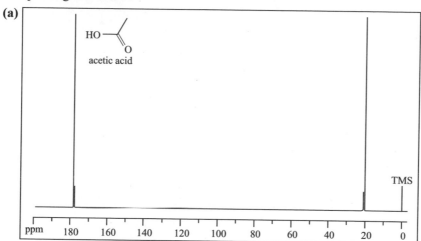

(b)

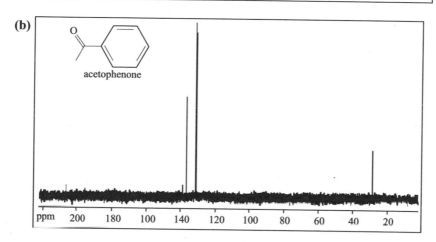

(c)

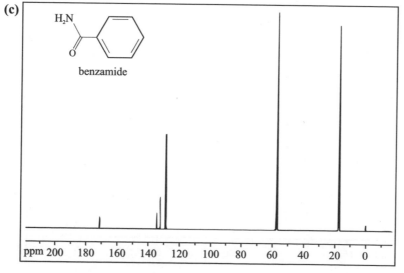

(d)

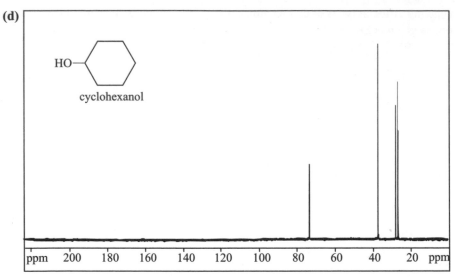

cyclohexanol

(e)

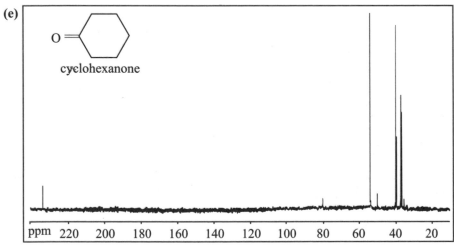

cyclohexanone

(f)

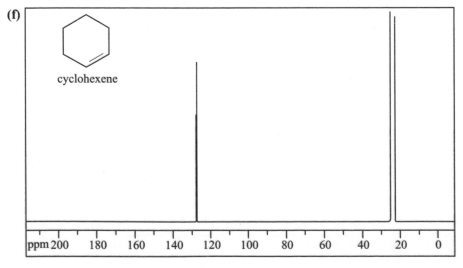

cyclohexene

(g)

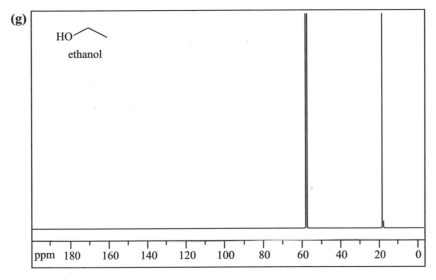

HO⌒ ethanol

(h)

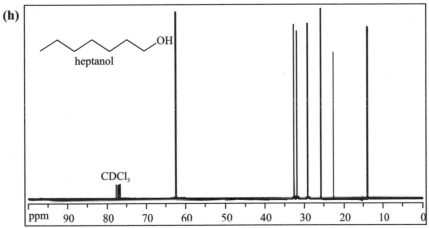

heptanol

CDCl₃

(i)

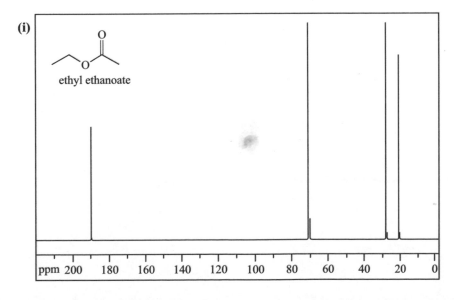

ethyl ethanoate

(j)

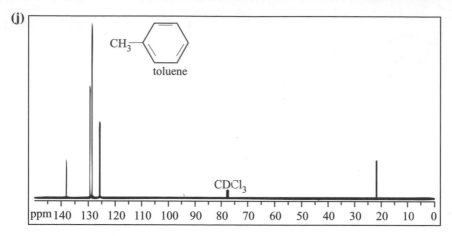

(k)

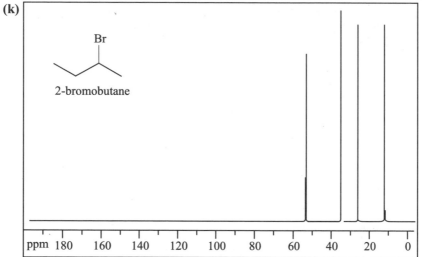

(l)

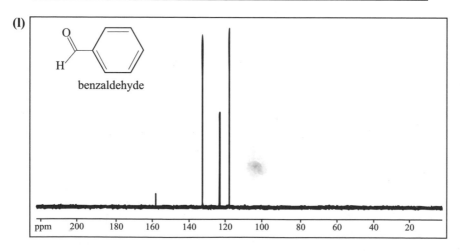

Mass Spectrometry

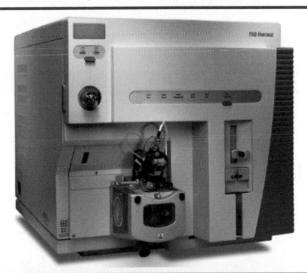

Francis William Aston (1 September 1877 – 20 November 1945) was an English chemist and physicist who won the 1922 Nobel Prize in Chemistry for his discovery, by means of his mass spectrograph.

Mass Spectrometry

6.1 INTRODUCTION

Mass spectrometry is a technique used to characterise the organic molecules by measuring its exact molecular weight and thus the exact molecular formula. The basic principle of mass spectrometry (MS) is to generate ions from either inorganic or organic compounds by any suitable method, to separate these ions by their mass-to-charge ratio (m/z) and to detect them qualitatively and quantitatively by their respective m/z and abundance. The mass spectrometer records the mass spectrum, which is the two-dimensional representation of relative abundance versus m/z of ionic species of that respective m/z ratio which have been created from the analyte within the ion source. The mass-to-charge ratio, m/z, is dimensionless, because it calculates from the dimensionless mass number, m, of a given ion, and the number of its elementary charges.

6.2 INSTRUMENTATION

A mass spectrometer consists of an ionisation chamber, mass analyzer and detector (**Figure 1**).

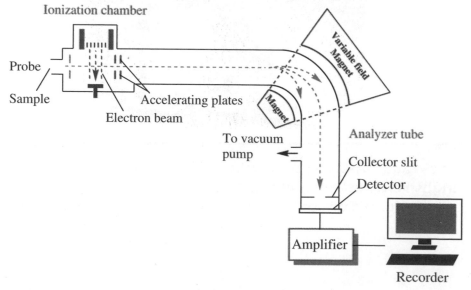

Figure 1. Diagramatic representation of mass spectrometer.

6.2.1 Ionisation Chamber (Ion Source)

In the source region, neutral sample molecules are ionised and then accelerated into the mass analyzer. The small quantity of sample under analysis is transferred from inlet to the ionisation chamber, where the pressure is greater than the ionization chamber. Energy of 10 eV is usually necessary to remove an electron from the parent molecule to give the molecular ion, while higher energy (70 eV) is required for the fragmentation of parent ion i.e., the molecular ion initially produced. The very low pressure is intended to prevent undue scattering of ions by collisions.

6.2.2 Mass Analyser

The mass analyser is the heart of the mass spectrometer. This section separates ions, either in space or in time, according to their mass to charge ratio. The positively charged ion (parent or fragment ions) produced in the ion chamber are accelerated by applying an acceleration potential. These ions then enter the mass analyzer. Here the fragment ions are differentiated on the basis of their m/z ratio. The positive ions travel through whole of the analyser portion of the mass spectrometer with high velocity are separated according to their m/z ratio.

6.2.3 Detector

After the ions are separated, they are detected and the signal is transferred to a data system for analysis. All mass spectrometers also have a vacuum system to maintain the low pressure, which is also called high vacuum, required for operation. High vacuum minimizes ion-molecule reactions, scattering, and neutralisation of the ions.

6.3 IONISATION TECHNIQUES

6.3.1 Electron Impact (EI)

Electron impact is a technique in which sample gets converted into gas phase by heating under vacuum followed by the bombardment of a stream of electrons in order to ionise the sample. These electrons are generated by the metallic filament and accelerated through a potential difference and their impact on the gaseous sample molecules results in the ionisation of these molecules.

6.3.2 Chemical Ionisation (CI)

Chemical Ionisation is closely related to EI, as it also uses the stream of electrons in ionisation process. However in CI, it is not the sample molecules which are ionised, but a reagent gas, usually ammonia or methane, which is present at much higher concentration. And the main difference is that the sample is ionised by a strong acid produced by the ionisation of the reagent gas.

6.3.3 Atmospheric Pressure Chemical Ionisation (APCI)

In APCI the same processes occur as in CI, but at atmospheric pressure. Through a similar mechanism the reagent gas becomes protonated and can act as an acid towards the sample, leading to the addition of a proton. The species formed in positive ion mode is $[M+H]^+$, while in negative ion mode the reagent gas acts as a base towards the sample and deprotonation occurs leading to the formation of $[M-H]^-$. APCI is also employed in LC-MS systems.

6.3.4 Fast Atom Bombardment (FAB)

FAB involves bombarding a solution of the analyte in a matrix with a beam of fast moving atoms, generally Xe atoms. The bombardment results in the transfer of energy from Xe atoms to the matrix, leads to the breaking of intermolecular bonds and the desorption of the sample into the gas phase. FAB has been widely used for the ionisation of large polar molecules.

6.3.5 Electrospray Ionisation (ESI)

ESI is a technique for the soft ionisation of a wide range of polar molecules (including biomolecules). The sample is usually dissolved in the organic solvents (commonly acetonitrile or methanol) and water with a pH modifier (formic or acetic acid), which ensures that ionisation takes place in the solution phase (**Figure 2**).

6.3.6 Matrix Assisted Laser Desorption Ionisation (MALDI)

It is similar in the principle to FAB except that the energy is transferred to the matrix from a laser beam and the matrix employed must therefore have a chromophore which absorbs at the wavelength of the laser. The matrix absorbs a pulse of energy from the laser beam and undergoes rapid heating, which ultimately leads to the vaporisation and ionisation of the sample molecules.

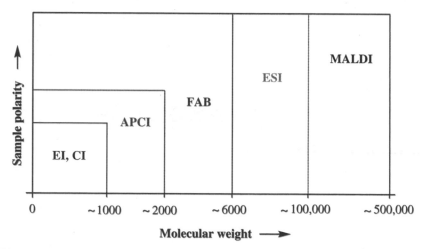

Figure 2. Range of molecular weight involved in different ionisation techniques.

═══════════ EXERCISE PROBLEMS 6.1 ═══════════

1. What is the basic principle of mass spectrometry?

2. What is the difference between spectrometry and spectroscopy?

3. What is the difference between electron impact (EI) and chemical ionisation (CI)?

4. What do you mean by atmospheric pressure chemical ionisation (APCI) in mass spectrometry?

5. What is fast atom bombardment (FAB) in mass spectrometry?

6. What is electrospray ionisation (ESI) in mass spectrometry?

7. What is matrix assisted laser desorption ionisation (MALDI) in mass spectrometry?

6.4 SOME USEFUL TERMS USED IN MASS SPECTROMETRY

6.4.1 Molecular Ion

The molecular ion peak is both an important reference point and integral in identifying an unknown compound. Although the molecular ion peak should be the most abundant peak in the spectrum, but this is not the case for the majority of compounds. Compounds like alcohols, nitrogen containing compounds, carboxylic acids, esters, and highly branched compounds may completely lack a visible molecular ion. The mass of the molecular ion is based upon the mass of the most abundant isotope for each element in the molecule. This is not the atomic weight from the periodic table.

6.4.2 Fragmentation

The structural information about an unknown sample cannot be obtained just by the molecular ion but it can be obtained by the fragmentation patterns of the mass spectrum. Fragmentation patterns are often complex, but provides important information to identify the structure of the molecule. However, all the fragment ions are, not of equal significance. Generally odd electron ions are formed *via* a rearrangement and have more significance than the even electron ions. Intensities of fragment ions in the mass spectrum depend on the energy relationships of the bonds broken and formed during reaction leading to the ion conditions of very low pressure and unimolecular reaction in a mass spectrometer. After a molecule is ionised, the molecular ion retains the excess ionisation energy. If this excess energy is greater than the energy required to break a chemical bond, the molecule can fragment. The fragmentation can be through direct cleavage or rearrangement.

6.4.2.1 Cleavage

It is the breaking of a bond to produce two fragments. These reactions generally produce an even electron ion (AB^+). The even electron ion is detected at an odd *m/z* value (assuming no nitrogen) and a neutral odd electron radical. Since the radical is a neutral fragment it is not observed in the mass spectrum.

6.4.2.2 Rearrangement

Rearrangements are the thermodynamically favorable complex reactions that involve both bond making and breaking. The rearrangement leads to the formation of odd electron ions that can be easily identified and these fragments often provide important information about the position and identity of functional groups.

6.4.3 Loss of Small Molecules

Small stable molecules such as H_2O, CO_2, CO and C_2H_4 can be lost from a molecular ion. For example, alcohol readily loses H_2O and shows a peak at 18 mass units less than the peak of molecular ion. This peak is referred as M-18 peak.

6.4.4 Isotope Abundance

Isotopic contributions: Many of the elements common to organic compounds occur naturally as isotopic mixtures, and these isotopes generally differ from each other by either 1 unit or by unit mass number. The natural abundance ratio of the element and their isotopes are in (**Table 1**). Mass spectrometers can separate these isotopes and measure their relative abundance. It provides useful information for identifying the elements in an ion. Since the majority of elements have two or more isotopes, the ratio of these isotopes can be a powerful tool in deriving the composition of unknown samples.

Table 1. Natural abundances and masses ($^{12}C = 12.000\ 000$) for common elements.

Isotope	Occurrence in nature (%)	Isotopic mass/mu
^1H	99.984	1.007 825
^2H	0.016	2.014 102
^{12}C	98.9	12.000 000
^{13}C	1.1	13.003 354
^{14}N	99.64	14.003 074
^{15}N	0.37	15.000 108
^{16}O	99.76	15.994 915
^{17}O	0.0317	16.999 133
^{18}O	0.2	17.999 160
^{19}F	100	18.998 405
^{28}Si	92.2	27.976 927
^{29}Si	4.7	28.976 491
^{30}Si	3.1	29.973 761
^{31}P	100	30.973 763
^{32}S	95.0	31.972 074
^{33}S	0.76	32.971 461
^{34}S	4.2	33.967 865
^{35}Cl	75.8	34.968 855
^{37}Cl	24.2	36.965 896
^{79}Br	50.5	78.918 348
^{81}Br	49.5	80.916 344
^{127}I	100	126.904 352

Since the natural abundance of the heavier isotope is generally much more less than that of the lightest isotope, the intensities of the isotope peaks are very low relative to the parent ion peak. Thus, M+1 isotope peak appearing due to ^{13}C is about 1.1% of the molecular ion peak because the natural abundance of ^{13}C isotope is about 1.1% as compared to 98.9% ^{12}C. The presence of ^2H will make an additional but very small (0.016%) contribution to the M+1 peak as its natural abundance is about 0.016% as compared to 99.98% of ^1H. Because of very low natural abundance of ^{13}C and ^2H their probability of finding in the molecule is so low that M+2 peaks due to ^{13}C and ^2H are often negligible. Thus, for the most of the organic compounds M+2 peaks are too small to be considered. However, for compounds, chlorine, bromine or sulphur, the M+2 isotope peak is important. The M+2 isotope peak due to chlorine atom is about 33% and 50% for the bromine atom of the molecular ion peak, respectively. Thus, in chloro or bromo compounds the ratio of the intensity of M$^+$ and M+2 peaks will be 3 : 1 or 1 : 1, respectively due to their natural abundance (**Figure 3**). Similarly, if one sulphur atom is present in a molecule, then according to the natural abundance of ^{34}S, the M+2 peak will be about 4.4% of the parent peak. Thus, the presence of Cl, Br or S can be easily detected on the basis of the intensity of M+2 peaks relative to the parent peak. Fluorine and iodine have no isotopes.

The molecular ion region is complex for those molecules containing more than one atom which has a significant isotope, e.g., C, Cl, S, Br. The relative intensities of the peaks in the molecular ion region for these molecules may be calculated from the expression:

$$(a + b)^m$$

Where, a = relative abundance of the lighter isotope

b = relative abundance of the heavier isotope

m = number of atoms of the elements present in the molecule.

Thus, when two atoms of the element are present, the expression becomes:

$$(a + b)^2 = a^2 + 2ab + b^2$$

Term 1   Term 2   Term 3

Term 1 gives the relative intensity of the peak containing isotopc 'a'

Term 2 gives the relative intensity of the peak containing isotopes 'a' and 'b'

Term 3 gives the relative intensity of the peak containing isotope 'b'

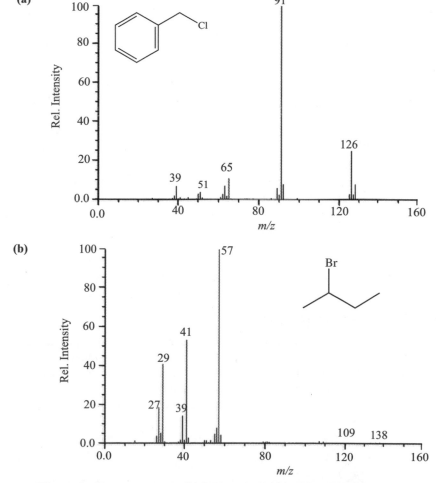

Figure 3. Mass spectrum of (a) Benzyl chloride; (b) 2-Bromobutane.

6.4.5 Metastable Ions

Metastable ions provide useful information on the nature of fragmentation process. Stable ions reach the detector without any fragmentation (k < 105 s^{-1}). They are produced from fragmentation that take place during the flight down in the ion tube and have low kinetic energy than normal ions. They have the same mass as normal ions, but less translational energy. They do not necessarily occur at integral value of m/z. The metastable peaks are broad due to some of the excitation energy leading to bond rupture may be converted to additional kinetic energy. For example, the broad peak at non-integer mass $m/z = 60.2$ and $m/z = 43.4$ is due to the metastable ions.

The presence of $m*$ ion in a mass spectrum indicates that the parent ion undergoes decomposition in one step to the daughter, so that it is of considerable mechanistic importance to investigate metastable ions. The m/z value of the metastable ion ($m*$) can be calculated by using the formula:

$$m* = (m_2)^2/m_1$$

Where, m_1 = mass of parent ion, and m_2 = mass of normal daughter ion.

6.4.6 Even-electron Rule

Odd-electron ions may eliminate either a radical or an even-electron neutral species, but even electron ions (such as protonated molecules or fragments formed by a single bond cleavage) will not usually loose a radical to form an odd-electron cation. Since the total energy of this product mixture would be too high, i.e., the successive loss of radicals is forbidden. A$^+$ ion will degrade to another ion and a neutral molecule.

$$A^+ = B^+ + C$$

Radical ions being odd electron species can extrude a neutral molecule leaving a radical ion as a co-product. Radical ion can also degrade to a radical and an ion.

$$D^{\bullet+} = E^{\bullet+} + F$$

6.4.7 Nitrogen Rule

Nitrogen rule is used to identify the type of molecular ion i.e., even or odd numbered m/z. If a compound contains an even number of nitrogen atoms (0, 2, 4, ...), its mono isotopic molecular ion will be detected at an even-numbered m/z (integer value). For example, CH_4 ($m/z = 16$); CH_3OH ($m/z = 32$); $CClF_3$ ($m/z = 94$); H_2NNH_2 ($m/z = 32$); $C_5H_5N_2$ ($m/z = 94$) give their molecular ion at even mass number. Similarly, an odd number of nitrogens (1, 3, 5, ...) is indicated by an odd-numbered m/z, e.g., NH_3 ($m/z = 17$); $C_2H_5NH_2$ ($m/z = 45$). This could be attributed to the fact that nitrogen has even atomic weight and an odd valence, whereas, all other elements encountered in the spectrometry have either an even mass and an even valence or odd mass or an odd valence. The inclusion of any common stable isotope, except ^{18}O alters the use of nitrogen rule.

6.4.8 McLafferty Rearrangement

The McLafferty rearrangement is often observed for carbonyl compounds that contain a linear alkyl chain containing γ-hydrogen. If this alkyl chain is long enough, a six-membered ring forms from the carbonyl oxygen to the hydrogen on the fourth carbon. This spacing allows the hydrogen to transfer to the carbonyl oxygen *via* a six membered ring. The McLafferty rearrangement is energetically favorable because it results in the loss of a neutral alkene and formation of a resonance stabilized radical cation (**Scheme 1**).

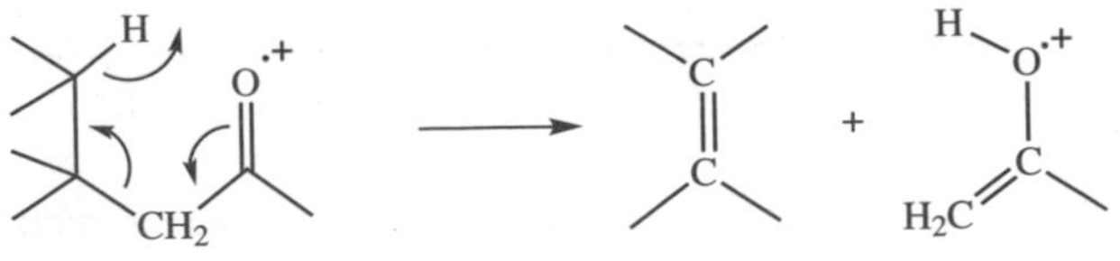

Scheme 1

6.5 APPLICATION OF MASS SPECTROMETRY

Ionisation of an organic molecule normally requires approximately 10–15 eV. In mass spectrometry, however, molecules are subjected to electronic bombardment of energy 70 eV. One electron is removed from the molecule, forming a high energy radical cation which has probability of fragmenting in order to dissipate its excess energy. As a rule, the electron will be lost from the most easily ionisable site in the molecule, e.g., from a lone pair of electrons on an atom such as O, N, S or halogen, or from unsaturated bond. If a molecule has no lone pairs or sites of unsaturation, then the electron will be lost from sigma bond. C–C bonds are more easily ionized than C–H sigma bonds.

6.5.1 Fragmentation Modes of Various Classes of Organic Compounds

6.5.1.1 Hydrocarbons

6.5.1.1.1 Alkanes

When saturated hydrocarbons are subjected to mass spectrometry, radical cations form. Since, there is no functional group in which to localise the charge, the charge may reside in any of the sigma bonds. Simple fission processes predominates, yielding odd m/z fragments.

6.5.1.1.1.1 Saturated hydrocarbon (alkanes) straight chain

The straight chain saturated hydrocarbons show molecular ion peak, but for long chain compounds, it is of low intensity.

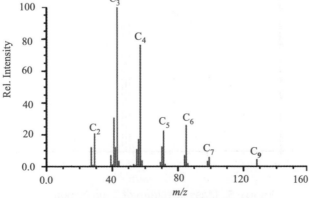

Figure 4. Mass spectrum of nonane.

In these compounds an alkyl group is lost from one end of the molecule, followed by successive losses of 14 mass units (CH_2). In the mass spectra of different compounds of homologous series, the height of the parent peak decreases as the molecular mass increases. In alkanes the most intense peaks are generally for C–3 and C–4 ions at m/z value of 43 and 57 respectively. Further the intensity decreases as the number of carbon atoms increases. The relative abundance of the carbocation, $viz.$, C_nH_{2n+1}, depends upon the stability of the ions and the radical lost. The peaks corresponding to C_nH_{2n+1} are also accompanied by $C_nH_{2n}^+$ and C_nH_{2n-1}. In case of molecular ion derived from n-alkanes an extensive decomposition takes place to give a series of fragment ions separated by 14 mass units (**Figure 4**).

6.5.1.1.1.2 Branched chain alkanes

The spectrum of the branched chain alkanes is nearly similar to those of straight chain compounds. Since, fragmentation lead to the formation of a more stable carbocation at each branch the gradual decrease of intensities is not observed. Generally, the relative abundance of the parent ion is least due to the preferable cleavage of the bond.

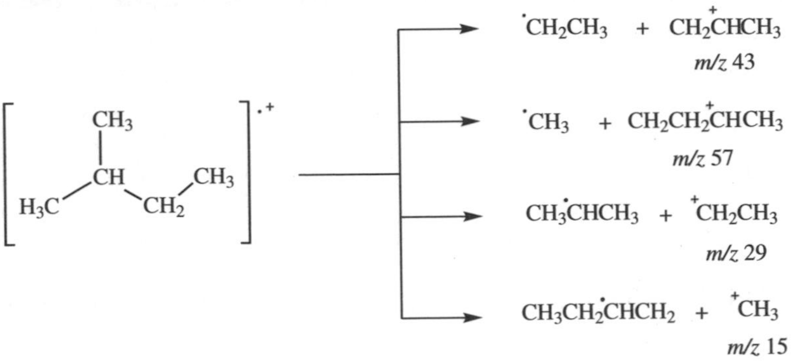

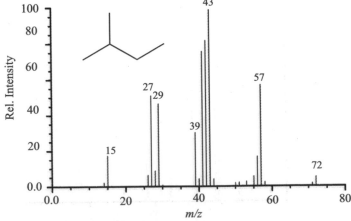

Scheme 2

Similar to the straight chain alkanes they also show the carbocation with C-3 and C-4 as intense peaks. The fragmentation at the branching point is often accompanied by hydrogen rearrangement leading to the C_nH_{2n} peak as more prominent and sometimes larger than C_nH_{2n+1} peak. As the alkane portion of any molecule becomes larger, the presence of the C_nH_{2n+1} peak becomes more prominent (**Scheme 2**).

Figure 5. Mass spectrum of 2-methylbutane.

6.5.1.1.1.3 Cyclic alkanes

In cyclic alkanes the relative abundance of the molecular ion is more as compared to the straight chain alkanes with same number of carbon atoms. If there is moiety attached to the ring, cleavage is preferred at the bond connecting the ring to the rest of the molecule. Generally, the fragmentation of the cyclic structure is caused by the loss of more than two carbon atoms. The loss of a methyl radical occurs less frequently because the methyl radical is less stable in comparison to the neutral ethylene molecule at M–28 or an ethyl radical at M–29. In cyclic alkanes the characteristic peaks are formed corresponding to C_nH_{2n-1} and C_nH_{2n-2} series (**Figure 6, Scheme 3**).

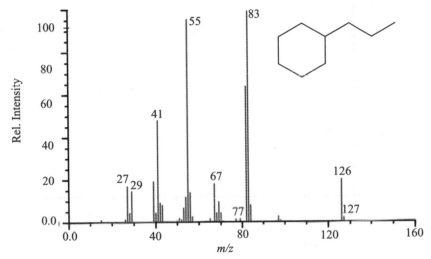

Figure 6. Mass spectrum propyl cyclohexane.

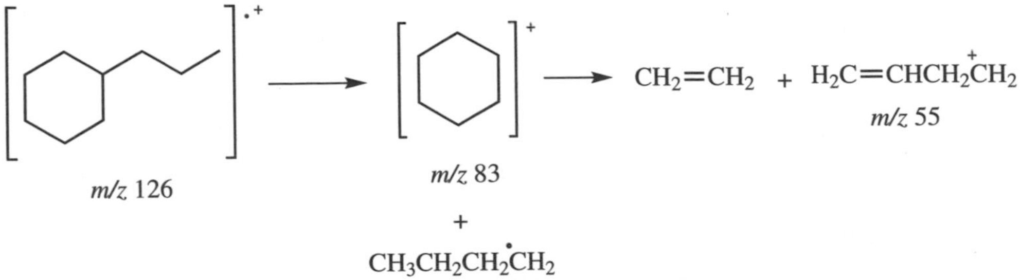

Scheme 3

6.5.1.1.2 Alkenes

The molecular ion peak of alkenes is more intense than the corresponding alkanes due to better resonance stabilisation of the charge on the cation formed from the olefin. Generally, the fragmentation takes place at allylic position. The mass spectrums are characterised by peaks at intervals of 14 mass units (CH_2), similar to saturated hydrocarbons and the more intense fragments (C_nH_{2n} ion) are formed by McLafferty rearrangement (**Scheme 4, Figure 7**).

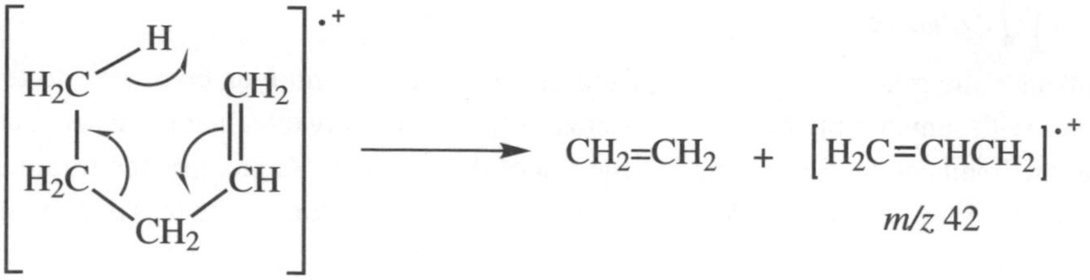

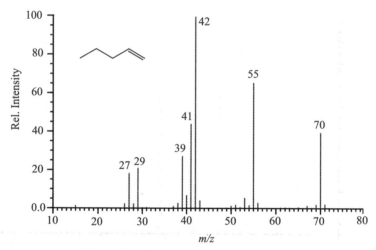

Scheme 4

Figure 7. Mass spectrum of 1-pentene.

6.5.1.1.2.1 Cyclic alkenes

They show a distinct molecular ion peak. A unique mode of cleavage is a type of *retro*-Diel's Alder reaction (**Scheme 5**).

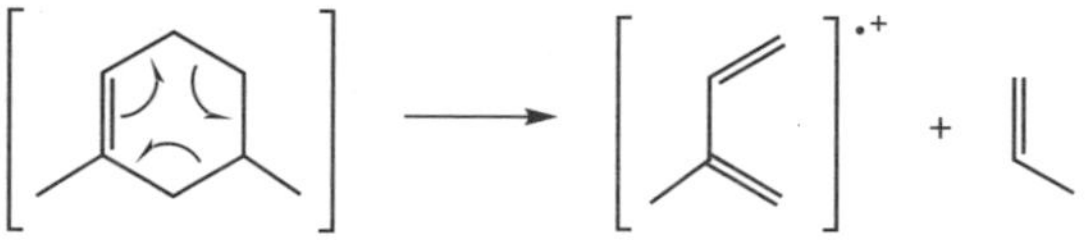

Scheme 5

6.5.1.1.3 Alkynes

The mass spectrum of alkynes generally consists of a distinct molecular ion peak. M–15, M–29, M–43 etc. 1-Pentane are common fragment ions which are generally formed by the loss of alkyl radicals. Alkynes having γ-hydrogen with respect to the triple bond undergoes the McLafferty rearrangement.

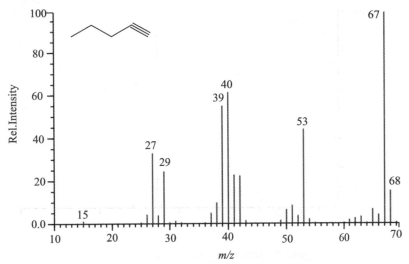

Figure 8. Mass spectrum of 1-pentyne.

6.5.1.1.4 Aromatic hydrocarbons

In aromatic hydrocarbons the molecular ion peak is the base peak due to its stabilisation through resonance. Generally, the alkyl benzene undergoes dominant fragmentation at the benzylic position to give molecular ion (m/z = 91) that rearranges to seven membered tropylium ion (**Scheme 6**).

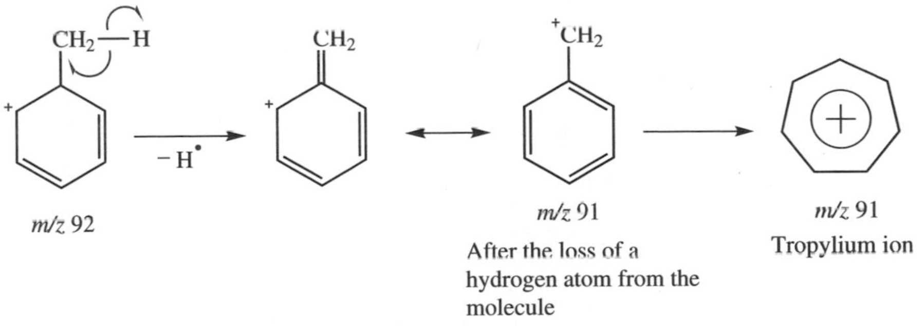

m/z 92

$-H^{•}$

m/z 91
After the loss of a
hydrogen atom from the
molecule

m/z 91
Tropylium ion

Scheme 6

Benzene rings with highly branched substituted groups produce fragments larger than m/z 91 by the intervals of 14 units. The largest of these peaks will result in a highly substituted cations and a large radical, like a simpler branched alkanes (**Figure 9**).

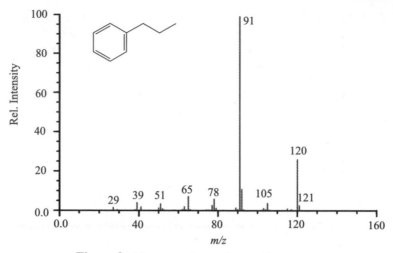

Figure 9. Mass spectrum of propyl benzene.

6.5.1.2 Alkyl halides

Alkyl halides undergo loss of $X^•$ leaving behind R^+. 2-Bromobutane (**Figure 3b**) undergoes fragmentation to give $m/z = 29, 41, 57, 56, 107$ (**Scheme 7**).

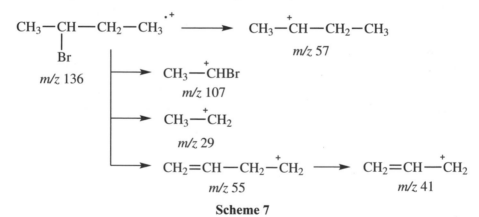

Scheme 7

In many cases alkyl halides do not exhibit a molecular ion peak, which may be due to the high probability of the loss of HX. Thus, the first possible peak in the mass spectrum of alkyl chloride could corresponds to M-64 (loss of HCl followed by $CH_2=CH_2$). This process is prominent as the number of carbon atom increases.

6.5.1.3 Hydroxyl compounds

6.5.1.3.1 Alcohols

6.5.1.3.1.1 Aliphatic Alcohols

The molecular ion peaks of primary and secondary alcohols are small, while that of tertiary alcohol is very weak. M-1 peak is formed due to the loss of single hydrogen atom from the α-carbon in primary and secondary alcohols. Alcohols undergoes β-cleavage to give resonance stabilized cations.

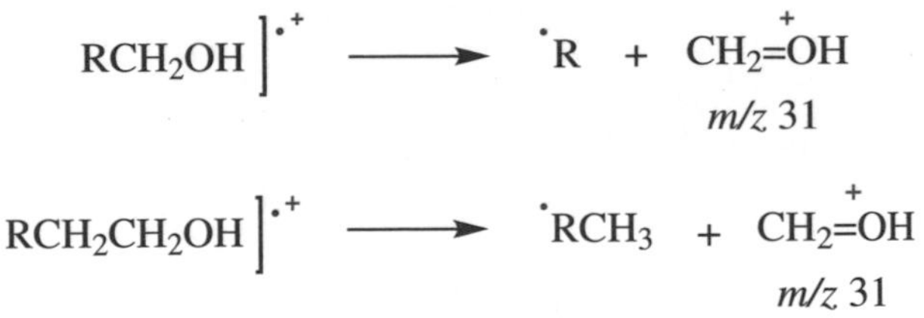

$$RCH_2OH \Big]^{\cdot +} \longrightarrow \ ^{\cdot}R \ + \ CH_2\overset{+}{=}OH$$
$$m/z \ 31$$

$$RCH_2CH_2OH \Big]^{\cdot +} \longrightarrow \ ^{\cdot}RCH_3 \ + \ CH_2\overset{+}{=}OH$$
$$m/z \ 31$$

Scheme 8

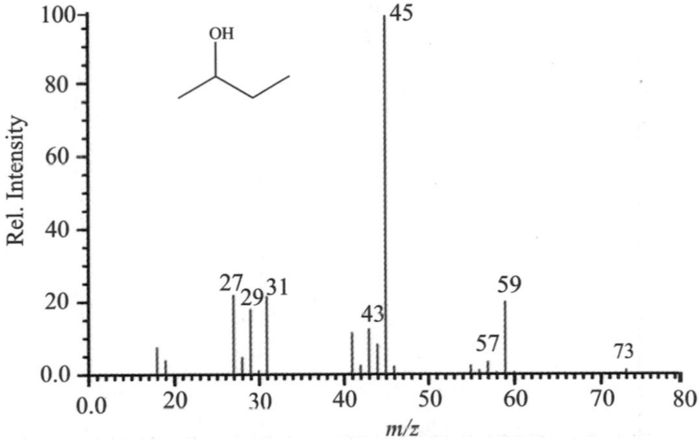

Figure 10. Mass spectrum of 2-butanol.

Thus, the primary alcohols show prominent peak due to $CH_2=OH^+$ ($m/z = 31$), secondary alcohols show peaks at 45, 59 and 73 etc., corresponding to $RCH=OH^+$ (**Figure 10**) while tertiary alcohols show peaks at 59, 73 & 87 etc., corresponding to $R_2C=OH^+$ (**Scheme 8**). Alcohols show distinct M-18 peak due to loss of water from the parent ion, where, M corresponds to molecular (i.e., parent) M–18 ion (**Scheme 9**). This is readily noticeable in the spectra of primary alcohols.

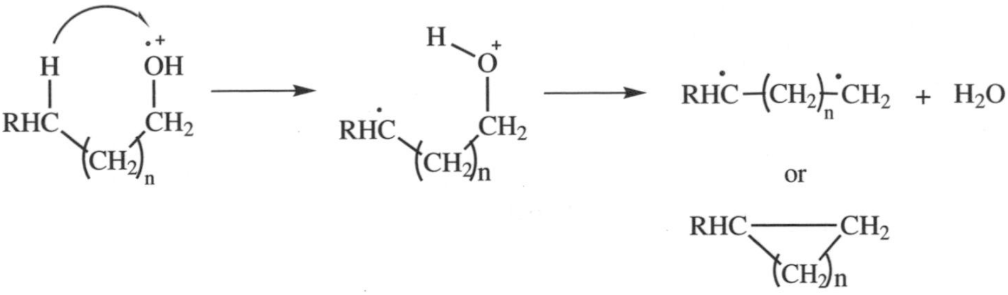

Scheme 9

6.5.1.3.1.2 Cyclic alcohols

In aromatic alcohols (e.g. benzyl alcohol), the parent ion is very prominent in the mass spectrum, due to stability of the benzylic cation (**Scheme 10, Figure 11**). They undergo complex fragmentation. Thus, cyclohexanol ($m/z = 100$) gives $C_6H_{11}O^+$ by simple loss of hydrogen; it loses H_2O to give $C_6H_9^+$ etc.

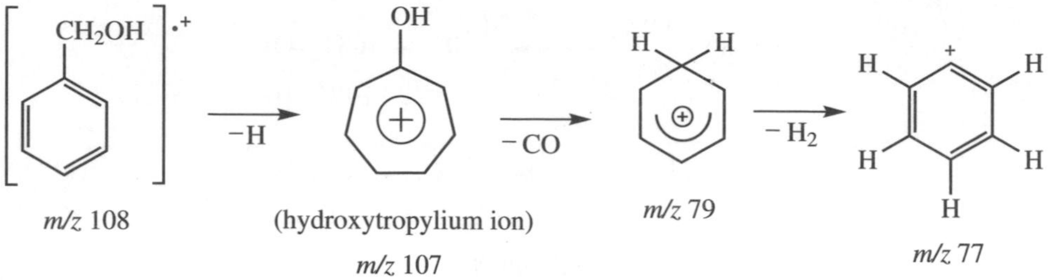

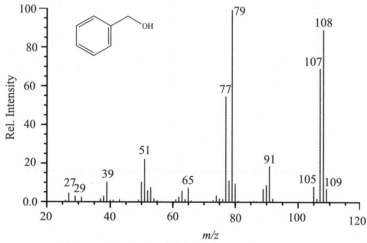

Scheme 10

Figure 11. Mass spectrum of benzyl alcohol.

Some ortho-substituted benzyl alcohol undergoes loss of H_2O to give distinct M–18 peak (**Scheme 11**, **Figure 12**).

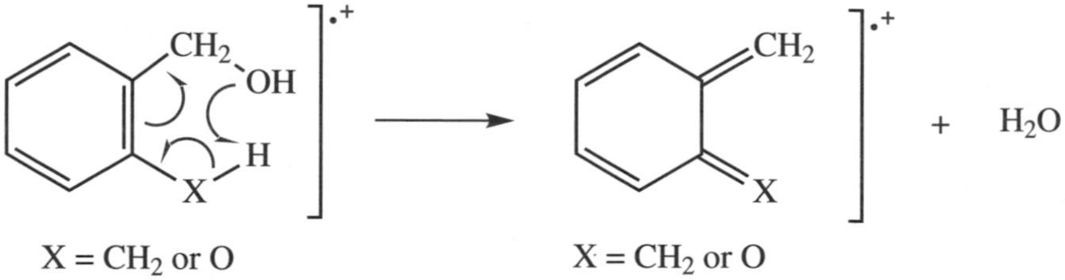

$X = CH_2$ or O $X = CH_2$ or O

Scheme 11

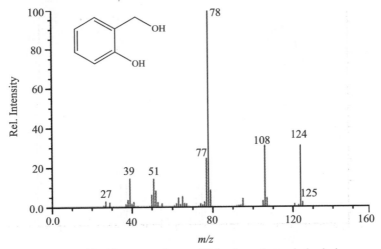

Figure 12. Mass spectrum of 2-hydroxy benzyl alcohol.

6.5.1.3.2 Phenols

Molecular ion peak is the base peak. M–28 peak is due to the loss of CO and M–29 peak is due to loss of CHO (**Scheme 12**).

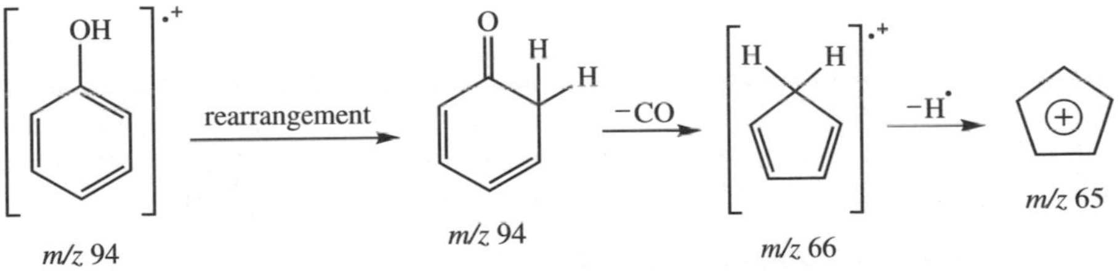

Scheme 12

Phenol with side chain undergoes benzylic cleavage to give hydroxyl tropylium ions, which further fragment to give $m/z = 77, 79$. The fragment $m/z = 77$ on elimination of a molecule of alkene yields $m/z = 51$, and C_3H_2 form $m/z = 39$ as other daughter ions (**Scheme 13, Figure 13**).

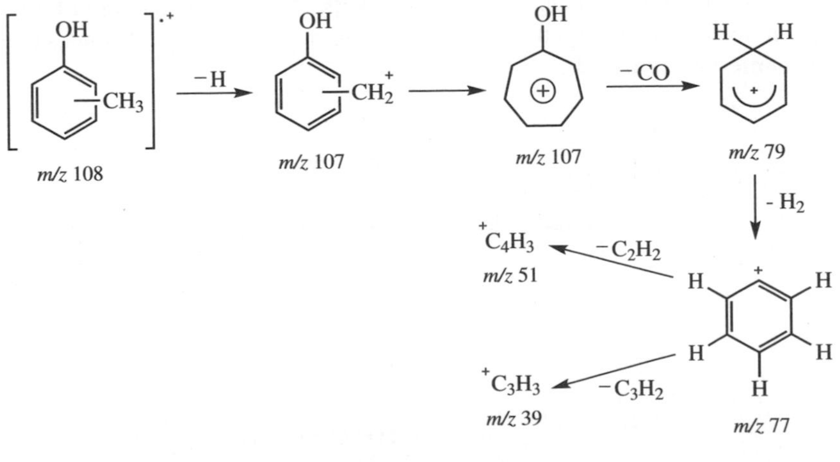

Scheme 13

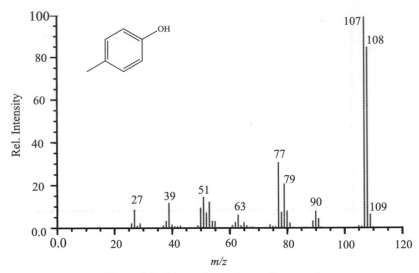

Figure 13. Mass of spectrum of *p*-cresol.

6.5.1.4 Aldehydes

6.5.1.4.1 Aliphatic aldehydes

6.5.1.4.1.1 α-Cleavage

The molecular ion peak of low intensity is observed. The major peaks observed in spectrum of aldehyde are the result of the α-cleavage to yield acylium ion, that further losses a R radical. This fragmentation results in an M-1 and M-R peak from the H–C=O$^+$ (m/z = 29) that are characteristic of aldehyde. The peak at m/z = 29 in higher aldehydes may also be due to $C_2H_5^+$.

6.5.1.4.1.2 β-Cleavage (McLafferty rearrangement)

The aldehydes that contain γ-hydrogen atom shows a prominent peak at m/z = 44 due to β-cleavage (McLafferty rearrangement). Straight chain aldehydes shows peaks at M–18, M–28, M–43, M–44, due to the loss of water, ethylene, CH_2=CH–O, CH_2=CH–OH.

6.5.1.4.2 Aromatic aldehydes

They exhibit a very intense molecular ion peak along with M-1 peak by the loss of hydrogen. The ArC≡O$^+$ fragment looses CO to form the phenyl ion at m/z = 77 that further degrades to give a peak at m/z = 51 (**Scheme 14, Figure 14**).

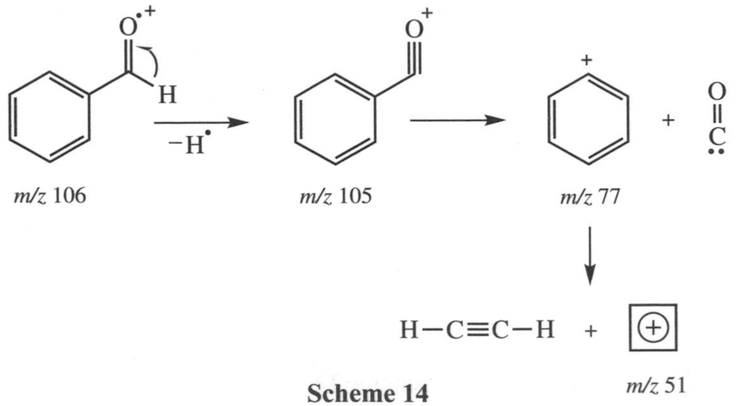

m/z 106 m/z 105 m/z 77

H–C≡C–H + ⊕

Scheme 14 m/z 51

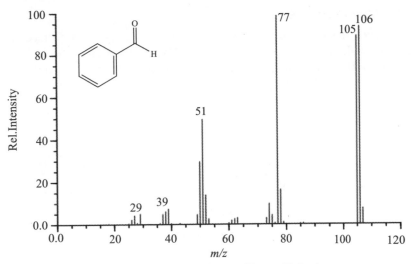

Figure 14. Mass spectrum of benzaldehyde.

6.5.1.5 Ketones

6.5.1.5.1 Aliphatic ketones

Aliphatic ketones show an intense molecular ion peak. Similar to alcohols and ethers, cleavage of the carbon-carbon bond next to the oxygen atoms takes place. The above cleavage gives $m/z = 43$ (where $R-CH_3$) or 57 or 71, etc. (**Figure 15**). The cleavage of the α-bond of ketones results in the resonance stabilised acylium ion. The base peak in the spectrum is usually formed by the removal of the larger alkyl group since it forms a more stable radical.

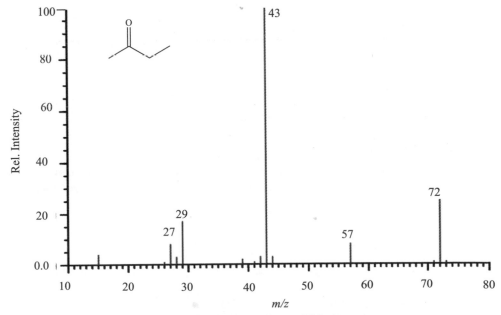

Figure 15. Mass spectrum of 2-butanone

6.5.1.5.2 Aromatic ketones

In aromatic ketones the molecular ion at $m/z = 105$ for $Ar-C=O^+$ is formed by the loss of R radical and is usually the base peak. Loss of CO from this ion gives the aryl ion at $m/z = 77$ (**Scheme 15**, **Figure 16**). Further fragmentation results in a peak at $m/z = 55$ after the loss of $HC\equiv CH$. In some aromatic ketones if the other alkyl component contains a γ-hydrogen atom they undergo the McLafferty rearrangement.

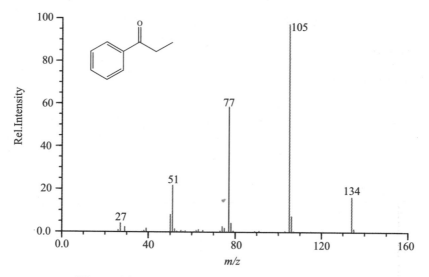

m/z 106 m/z 105 m/z 77

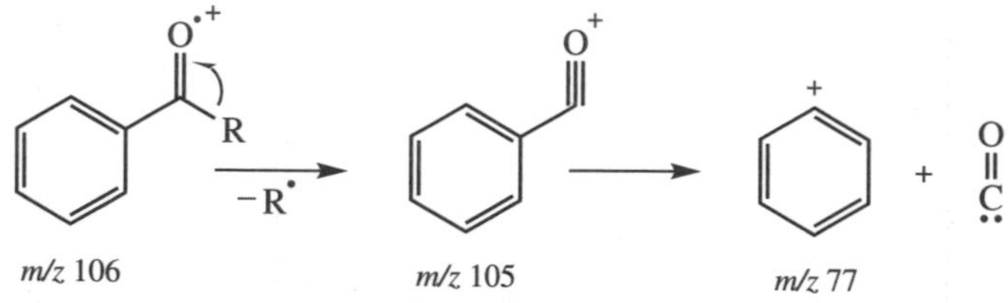

Scheme 15

Figure 16. Mass spectrum of phenyl-1-propanone.

6.5.1.6 Carboxylic acids, esters and amides

6.5.1.6.1 Aliphatic carboxylic acids

In aliphatic carboxylic acids, the molecular ion peak is weak. The carboxylic acids containing γ-hydrogen shows a prominent peak at $m/z = 60$ due to McLafferty rearrangement (**Scheme 16**, **Figure 17**).

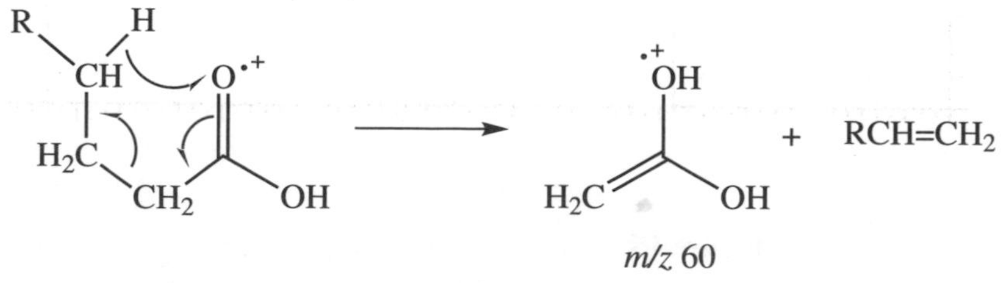

m/z 60

Scheme 16

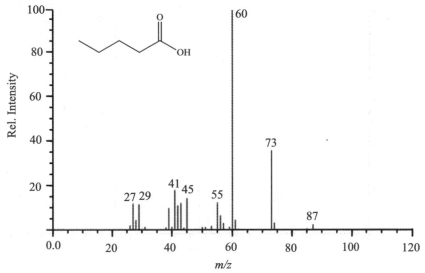

Figure 17. Mass spectrum of pentanoic acid.

In short chain carboxylic acids the peak at M–OH and M–COOH are prominent. In larger carboxylic acids these peaks are less prevalent. Long chain carboxylic acids consist of the fragments at $C_nH_{2n-1}O_2$ and the presence of the hydrocarbon fragment at C_nH_{2n+1} (**Scheme 17**).

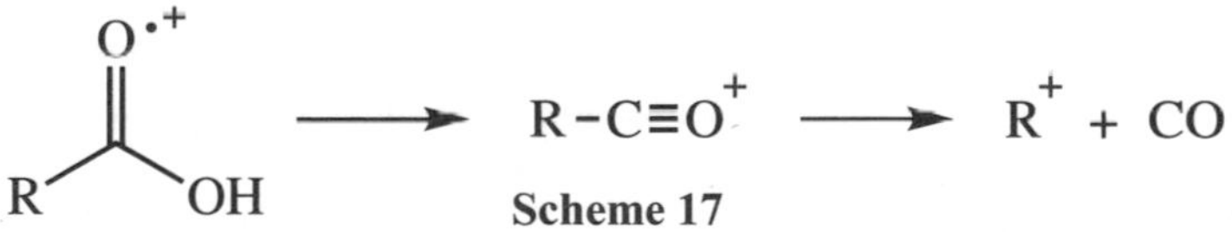

$$R-C\equiv O^+ \longrightarrow R^+ + CO$$

Scheme 17

6.5.1.6.2 Aromatic carboxylic acids

In aromatic acids the molecular ion peak is prominent. They produce large peaks at M–OH, M–CO$_2$H (**Scheme 18, Figure 18**).

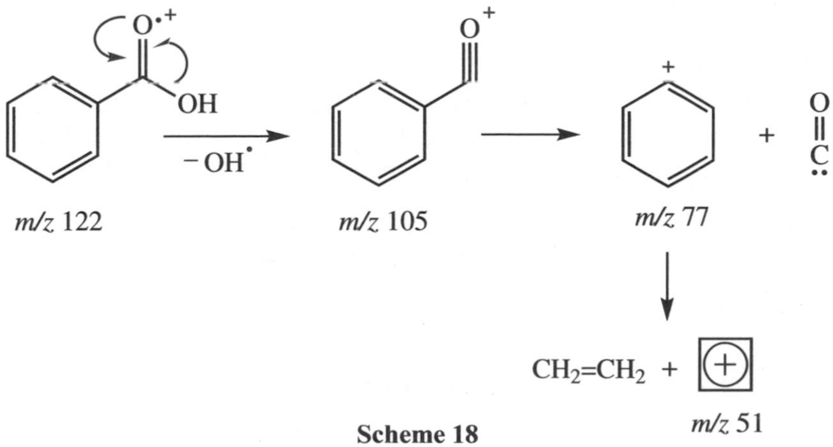

m/z 122 m/z 105 m/z 77

$$CH_2=CH_2 + \boxed{+}$$

Scheme 18 m/z 51

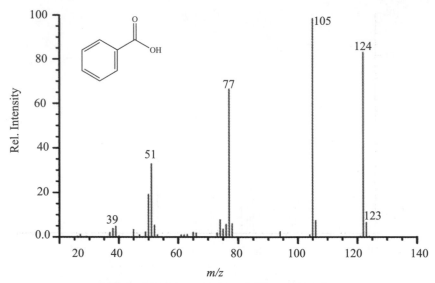

Figure 18. Mass spectrum of benzoic acid.

Some *ortho*-substituted aromatic carboxylic acids also loose water if an *ortho* group contains an abstractable hydrogen atom to give distinct M-18 peak (**Scheme 19**).

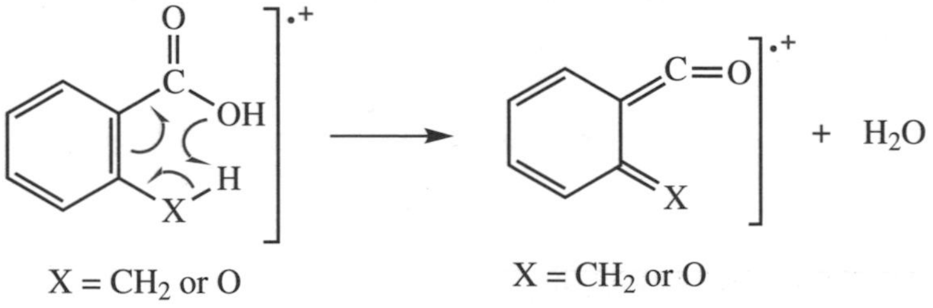

X = CH$_2$ or O   X = CH$_2$ or O

Scheme 19

6.5.1.7 Fragmentation of ethers

6.5.1.7.1 Aliphatic ethers

The molecular ion peak is usually weak in ethers. β-cleavage results in a resonance stabilised cation, which is prominent and sometimes the base peak. When the α-carbon is substituted the fragment can also undergo a rearrangement resulting in the base peak (**Scheme 20, Figure 19**).

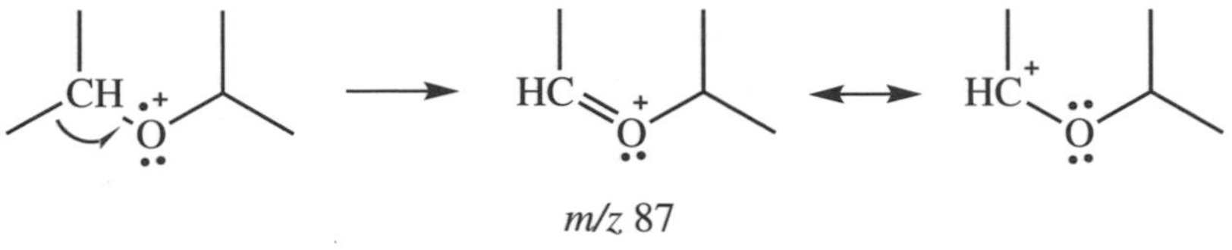

m/z 87

Scheme 20

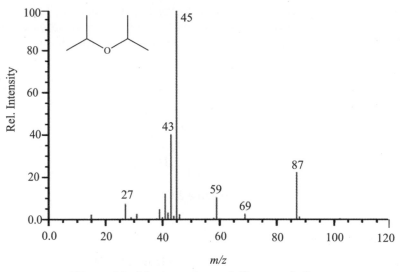

Figure 19. Mass spectrum of diisopropyl ether.

6.5.1.7.2 Aromatic ethers

A prominent molecular ion is observed with aromatic ether due to resonance stabilisation of the benzene ring. The major fractionation occurs at the β-bond to the aromatic ring, which can decompose further with the loss of CO (**Scheme 21, Figure 20**).

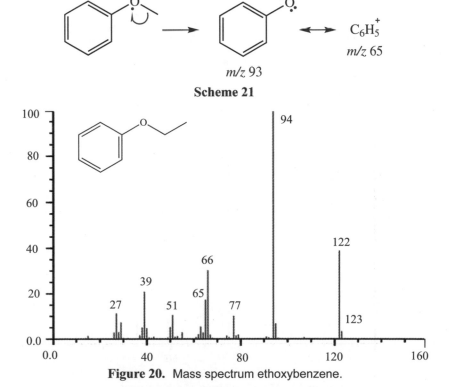

Figure 20. Mass spectrum ethoxybenzene.

They also undergo cleavage at the bond α to the ring to yield $m/z = 78$ and 77 due to hydrogen migration. When the alkyl chain is larger than two carbons, the β-cleavage is accompanied by hydrogen migration caused by the presence of the aromatic group. This cleavage results in a peak at $m/z = 94$ (**Scheme 22**).

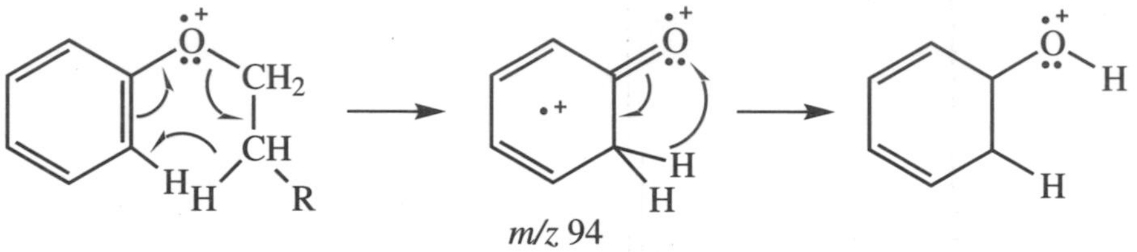

Scheme 22

6.5.1.8 Esters

6.5.1.8.1 Aliphatic esters

The molecular ion peak is apparent in aliphatic esters (**Figure 21**). The fragment due to α-cleavage usually observed. The resonance R–C=O⁺ stabilised ion gives a distinct peak for almost all esters. The R⁺ ion is prominent in short chain esters but is barely visible in esters with more than six carbon atoms. The ion R′COO⁺ distinct, if R′=CH₃. Since esters have both an alcohol and an acid component, therefore, it shows the fragmentation pattern similar to both of these types of compounds. The prevalence of the fragments is dependent on the size of each part of the ester. When the alcohol portion of the ester is prominent it fragments similar to that of an alcohol and looses a molecule of acid. Esters of long chain alcohols display peaks at 61, 75, 89 (**Scheme 23**). When the acid portion is the major component, the fractionation pattern is similar to carboxylic acids.

Scheme 23

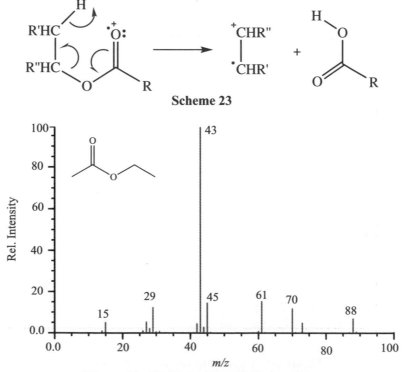

Figure 21. Mass spectrum of ethyl acetate.

6.5.1.8.2 Aromatic esters

In aromatic esters, the molecular ion peak of methyl esters of aromatic acids is prominent. The fragmentation leads to the generation of a base peak which is formed by the elimination of OR group while elimination of –COOR gives another prominent peak (**Scheme 24, Figure 22**).

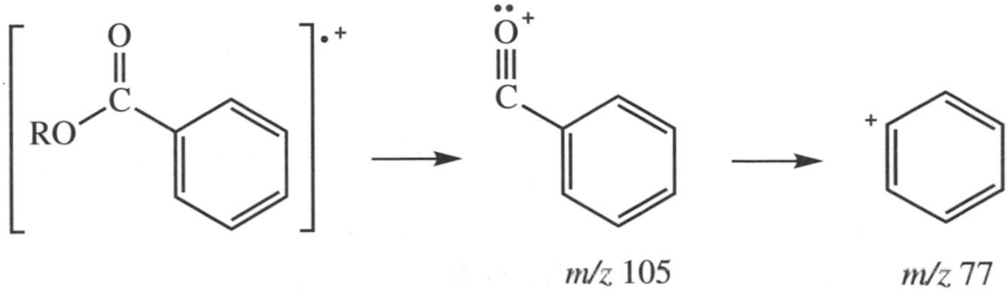

m/z 105   *m/z* 77

Scheme 24

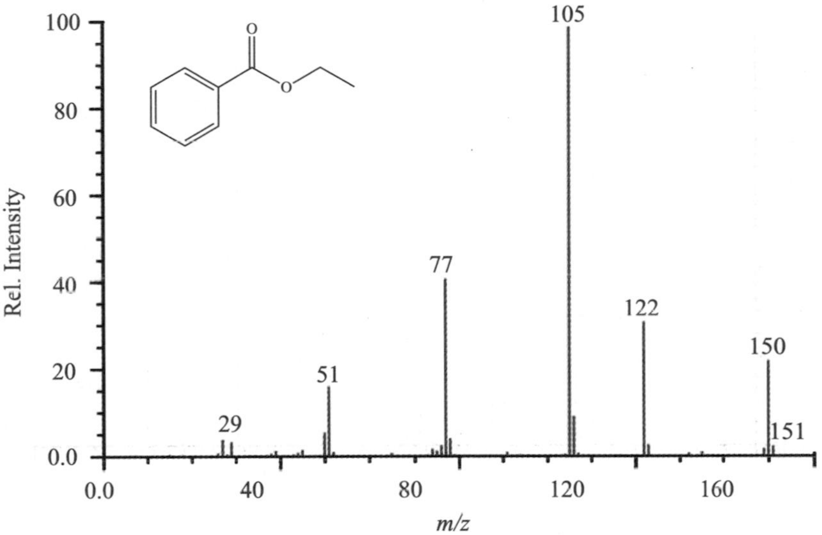

Figure 22. Mass spectrum of ethyl benzoate

6.5.1.9 Amines

6.5.1.9.1 Aliphatic amines

The fragmentation pattern is similar to that of alcohols and ethers. The base peak in most amines results from the β-cleavage. The largest alkyl group is lost as a radical in this cleavage (**Scheme 26**).

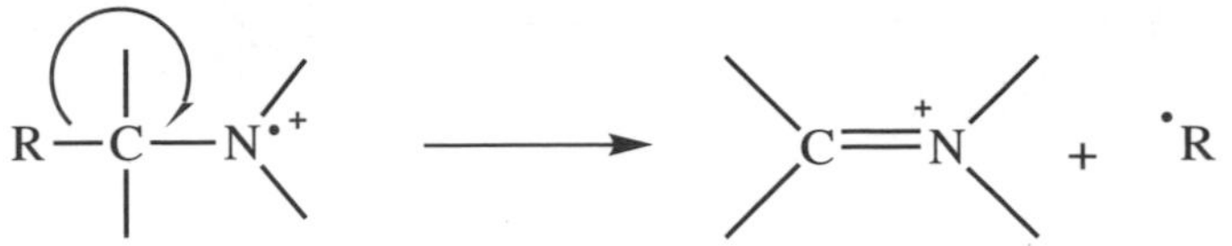

Scheme 25

    In case of all primary amines with an unbranched α-carbon, β-cleavage produces a peak at $m/z = 30$. The primary fragment from secondary or tertiary amine undergoes fragmentation *via* hydrogen rearrangement similar to aliphatic ethers to give a peak at $m/z = 30, 44, 58$ or 72 (**Scheme 25, Figure 23**).

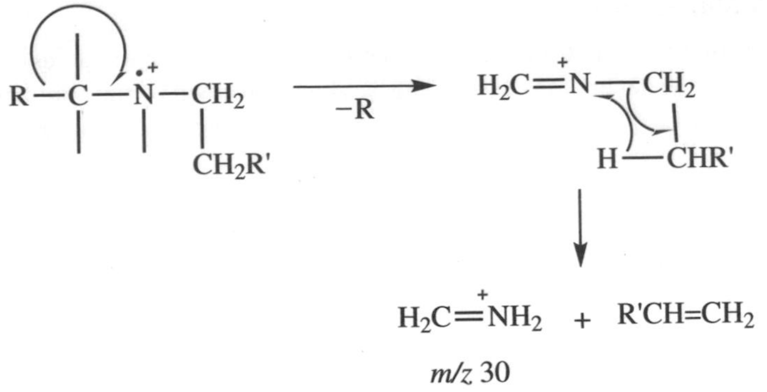

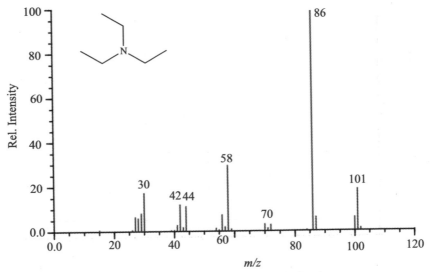

Scheme 26

Figure 23. Mass spectrum of triethylamine.

Longer chain amines give cyclic fragments. Primary straight chain amines show homologues series of peaks at $m/z = 30, 44, 58$............. resulting from the cleavage at C–C bonds successively removed from the nitrogen atom with retention of the charge on nitrogen containing fragment (**Scheme 27, Figure 24**).

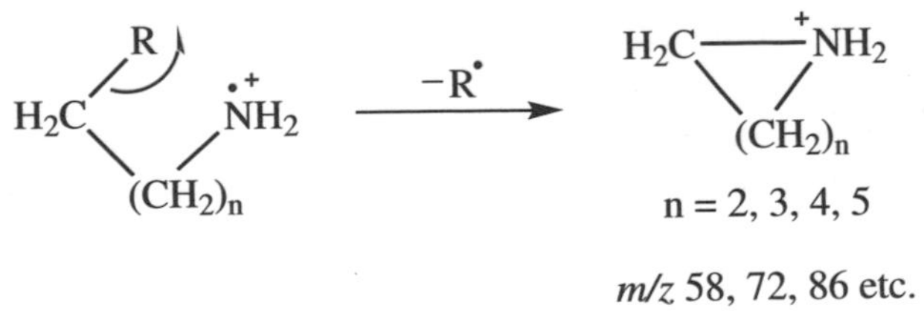

$$n = 2, 3, 4, 5$$

$$m/z\ 58, 72, 86\ \text{etc.}$$

Scheme 27

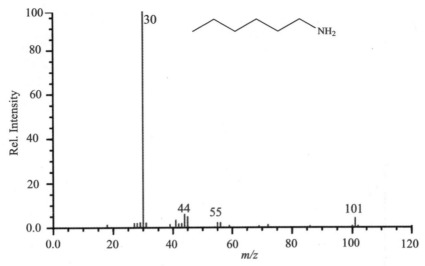

Figure 24. Mass spectrum of 1-hexylamine.

6.5.1.9.2 Cyclic amines

Cyclic amines usually show intense molecular ion peaks. The loss of an α-hydrogen atom leads to a strong M-1 peak. The ring undergoes β-cleavage followed by elimination of ethylene to give $CH_2–NH^+=CH_2$, $m/z = 43$ (**Figure 25**).

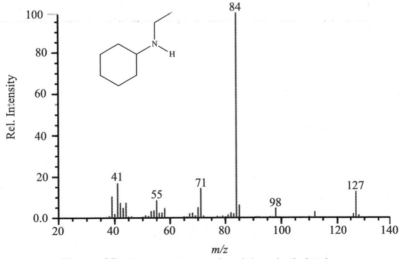

Figure 25. Mass spectrum of cyclohexyl ethylamine.

6.5.1.9.3 Aromatic amines

They show intense molecular ion peak and loss of one hydrogen atom of aniline gives a prominent M-1 peak. Loss of HCN followed by loss of a H-atom produces peaks at C_5H_6 and C_5H_5 (**Scheme 28**).

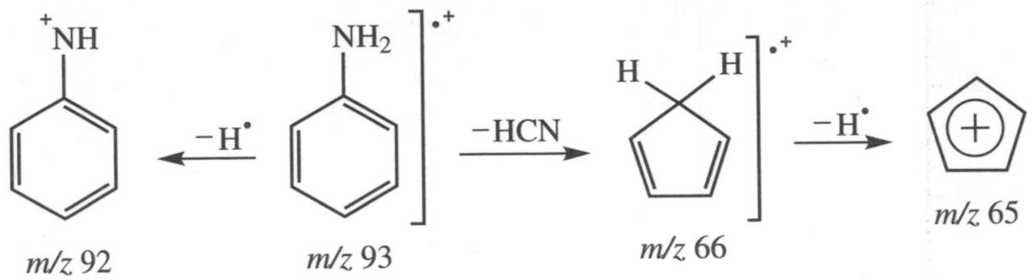

Scheme 28

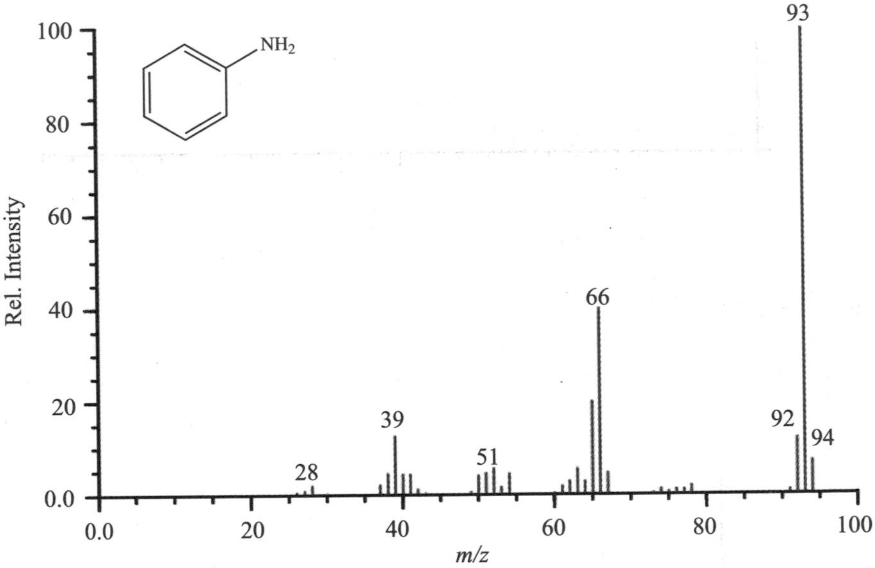

Figure 26. Mass spectrum of aniline.

6.5.1.10 Amides

6.5.1.10.1 Aliphatic amides

The molecular ion of most straight chain amides is apparently visible. The fractionation pattern is dependent on the length of the alkyl chain and the degree of substitution of the nitrogen group. Primary amides give a prominent peak from the McLafferty rearrangement (**Scheme 29**, **Figure 27**).

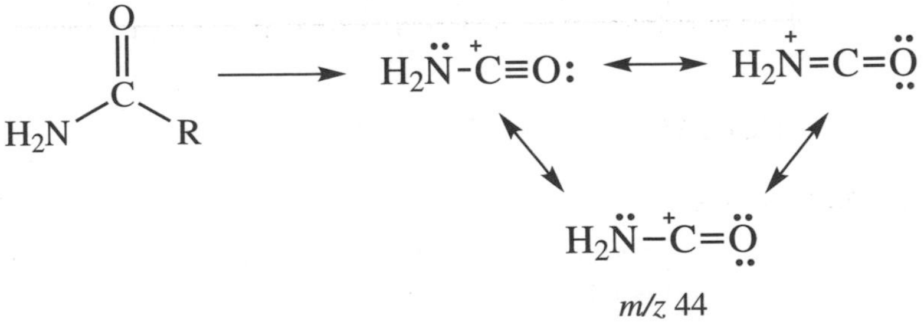

$$m/z \ 44$$

Scheme 29

The secondary and tertiary amides show the base peak at $m/z = 59$ due to McLafferty rearrangement.

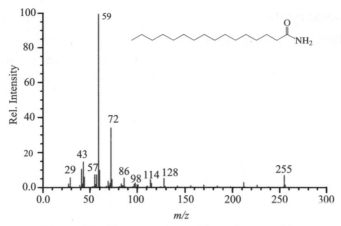

Figure 27. Mass spectrum of hexadecanamide.

Primary amines also produce a peak at $m/z = 86$ as a result of the following cleavage (**Scheme 30**).

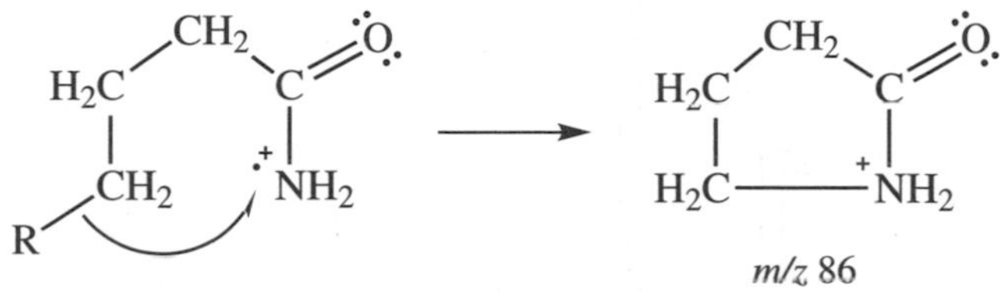

$$m/z\ 86$$

Scheme 30

6.5.1.10.2 Aromatic amides

Aromatic amides have a more prominent molecular ion peak. The loss of NR_2 generates a molecule of a resonance stabilised cation that subsequently losses a molecule of CO (**Scheme 31**, **Figure 28**).

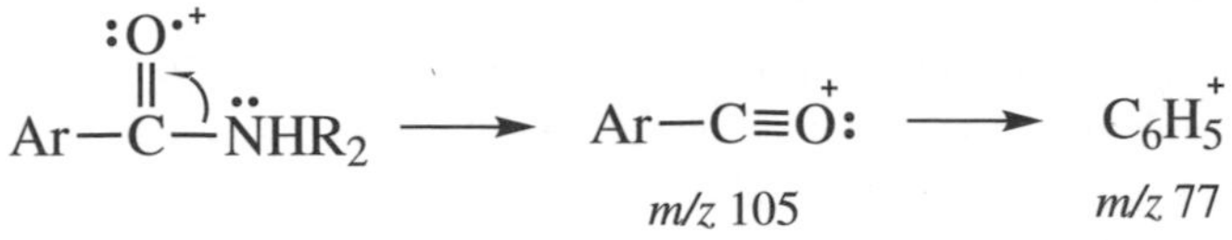

$$Ar-\overset{\overset{\displaystyle :O^{\cdot +}}{\|}}{C}-\overset{\displaystyle ..}{N}HR_2 \longrightarrow Ar-C\equiv\overset{\displaystyle +}{O}: \longrightarrow C_6H_5^+$$

$$m/z\ 105 \qquad\qquad m/z\ 77$$

Scheme 31

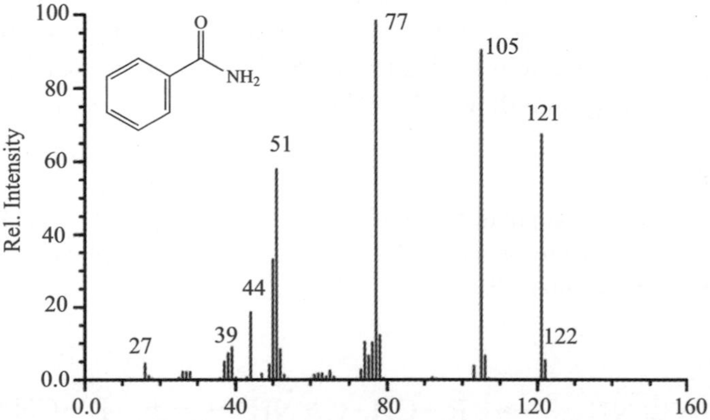

Figure 28. Mass spectrum of benzamide.

6.5.1.11 Nitro compounds

6.5.1.11.1 Aliphatic nitro compounds

They show very weak molecular ion peak (except in lower homologs). They undergo fragmentation to form hydrocarbon ions. The presence of a nitro group is indicated by an intense peak at $m/z = 30$ (NO^+) and a much smaller peak at $m/z = 46$ (NO_2^+). Minor peaks corresponding to the loss of hydroxyl radical and water molecule have been observed in the mass spectra of nitro compounds (**Figure 29**).

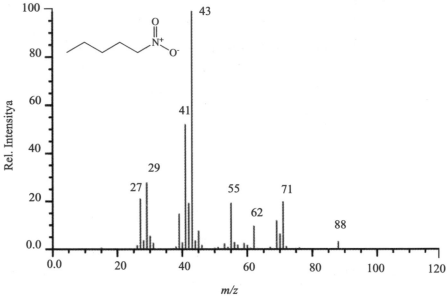

Figure 29. Mass spectrum of 1-nitropentane.

6.5.1.11.2 Aromatic nitro compounds

They show an intense molecular ion peak. The molecular ions rearrange to nitrites which readily lose nitric oxide to give aryloxy cation, followed by the loss of carbon monoxide. They show a small but distinct peak of M–O$^+$ ion. Loss of NO$_2$ radical from the molecular ion gives Ar$^+$ as the base peak, especially if this is stabilised by M$^+$ group or can rearrange into a tropylium ion. The mass spectrum of nitro compounds consists of M–16, M–30 and M–46 corresponding to M–O, M–NO, M–NO$_2$ ions. When the substituent is present in the *m*, *p*-positions, for example *m*- and *p*-nitroanilines, the fragmentation pattern similar to that of nitrobenzene is observed, while a hydrogen-containing substituent *ortho* to the nitro group shows M–OH peak.

6.5.1.12 Nitriles

The nitrile group containing compounds do not have a molecular ion. M-1 peak is formed by the loss of an α-hydrogen to form a resonance stabilised cation, which complicates the identification of molecular ion (**Scheme 32, Figure 30**).

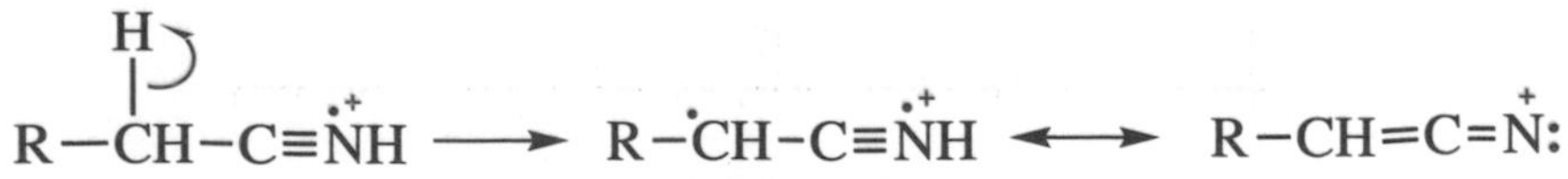

$$R-\overset{\overset{\displaystyle H}{|}}{C}H-C\equiv\overset{+}{N}H \longrightarrow R-\overset{.}{C}H-C\equiv\overset{+}{N}H \longleftrightarrow R-CH=C=\overset{+}{N}:$$

Scheme 32

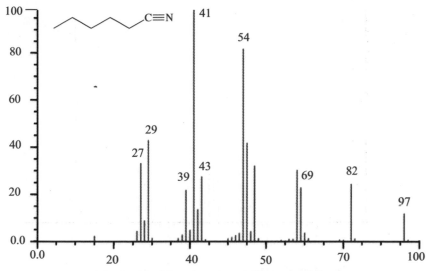

Figure 30. Mass spectrum of hexane nitrile.

6.5.1.13 Heterocyclic compounds

Aromatic heterocyclic compounds show intense molecular ion peak. They undergo fragmentation similar to benzene, e.g., molecular ion of benzene eliminates C_2H_2, whereas, pyrrole and pyridine loose HCN (**Figure 31a** and **31b**). Similarly, thiophene eliminates CHS radical and furan CHO radical from their parent ion. Pyrrole, thiophene and furan also eliminate CH_3 radical from their molecular ion to give $HC \equiv N^+$, $HC \equiv S^+$ and $HC \equiv O^+$ ions respectively. In alkyl substituted heteroaromatics, cleavage of the bond β to the ring is favoured, similar to alkylbenzenes. The fragment ions thus formed undergo ring expansion as benzyl cation changes to tropylium ion. This process is followed by the loss of HCN in nitrogen heterocycles.

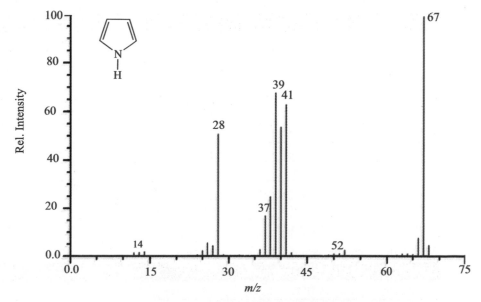

Figure 31a. Mass spectrum of pyrrole.

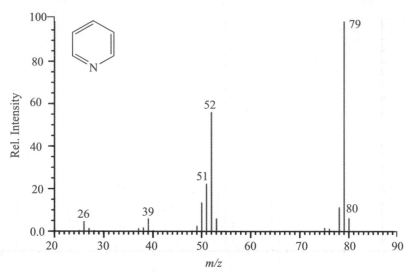

Figure 31b. Mass spectrum of pyridine

6.6 HOW TO ANALYSE MASS SPECTRA

Step 1: Analysis of molecular ion region.

    (a) Intensity of molecular ion?

    (b) Is 'M' odd or even?

    (c) Is isotope pattern significant?

    (d) What formula can be assigned? How many units of unsaturation (multiple bonds or rings) are indicated by each formula?

Step 2: Analysis of the fragment ions.

    (a) Are any characteristics losses apparent?

S.No.	Ion	Fragment	Structural or fragmentation types indicated
1	M-1	H	Aldehyde (Some ethers and amines)
2	M-15	CH_3	Methyl substituents
3	M-18	H_2O	Alcohols
4	M-28	C_2H_4, CO, N_2	C_2H_4 McLafferty rearrangement, CO, (extrusion from cyclic ketones)
5	M-29	CHO, C_2H_5	Aldehydes, ethyl substituents
6	M-34	H_2S	Thiols
7	M-35, M-36	Cl, HCl	Chlorides
8	M-43	CH_3CO, C_3H_7	Methyl ketones, propyl substituents
9	M-45	COOH	Carboxylic acids
10	M-60	CH_3COOH	Acetates

    (b) What formula can be assigned to significant ions?

S.No.	Ions	Fragments	Structural types indicated
1	29	CHO	Aldehyde
2	30	CH_2NH_2	Primary amines

S.No.	Ions	Fragments	Structural types indicated
3	43	CH_3CO, C_3H_7	Propyl substituents
4	29, 43, 57, 71	C_2H_5, C_3H_7, etc.	n-Alkyl
5	39, 50, 51, 52, 65, 77	Aromatic fragmentation products	Aromatic: most of these substituents will be present if an aromatic system is part of the structure
6	60	CH_3COOH	Carboxylic acids, acetates, methyl esters
7	91	$C_6H_5CH_2$	Benzyl
8	105	C_6H_5CO	Benzoyl

(c) Does the odd-even character of significant ions suggest a rearrangement process, i.e., Mclafferty, retro-Diel's Alder?

(d) Does high resolution analysis allow alternative formulas to be written for the ions'?

Step 3. Listing of partial structural units

(a) What are the partial structural units indicate?

(b) Are there relations between major ions (metastable peaks)?

(c) How many units of unsaturation and atoms are accounted for by the partial structural units? What residual fragments are possible?

Step 4. Postulating structures

(a) Combine partial and residual structures in all possible ways.

(b) Can any structure be eliminated on the basis of mass spectra or other data?

SOLVED PROBLEMS

Q1. Why fluorine and iodine containing compounds shows only molecular ion peak in mass spectrometry.

Sol. This is due to the reason that fluorine and iodine have no isotopes.

Q2. Why in case of n-alkanes a series of fragment ions separated by 14 mass units is observed?

Sol. This is due to the reason that in n-alkanes an extensive decomposition takes place to give a series of fragment ions separated by 14 mass units.

Q3. Why the molecular ion peak of alkenes is more intense than the corresponding alkanes?

Sol. Due to better resonance stabilisation of the charge on the cation formed from the olefin.

Q4. Why the mass spectrum of alkynes generally consists of distinct molecular ion peaks as M-15, M-29, M-43?

Sol. These are common fragment ions which are generally formed by the loss of alkyl radicals.

Q5. Why in aromatic hydrocarbons the molecular ion peak is the base peak?

Sol. Due to its stabilisation through resonance.

Q6. Why alcohols show distinct M-18 peak?

Sol. Due to loss of water from the parent ion, where M corresponds to molecular ion.

Q7. Why in benzyl alcohols, the parent ion is very prominent in the mass spectrum?

Sol. Due to stabilisation of the benzylic cation through resonance.

Q8. Explain, why aldehydes containing γ-hydrogen atom shows a prominent peak at $m/z = 44$.

Sol. Due to β-cleavage (McLafferty rearrangement).

Q9. Why does the organic compound containing even number of nitrogen shows the molecular ion peak at an even numbered m/z?

Sol. If a compound contains an even number of nitrogen atoms (0, 2, 4, ...), its mono isotopic molecular ion will be detected at an even-numbered m/z (integer value). This could be explained on the basis of nitrogen rule, which is based on the fact that nitrogen has even atomic weight and an odd valence, whereas, all other elements encountered in the spectrometry have either an even mass and an even valence or odd mass or an odd valence.

Q10. In aniline the following fragments are observed at $m/z = 92, 66, 65$ in mass spectrometry. Identify the structure.

Sol. Aniline undergoes fragmentation and results in fragments at $m/z = 92, 65$ due to the loss of H radical and loss of HCN molecule leads to $m/z = 65$.

Q11. (a) What type of hydrocarbon will give a spectrum with peak pattern shown below?

(b) What is the formula of hydrocarbon?

(c) What are the differences in mass between the major peaks?

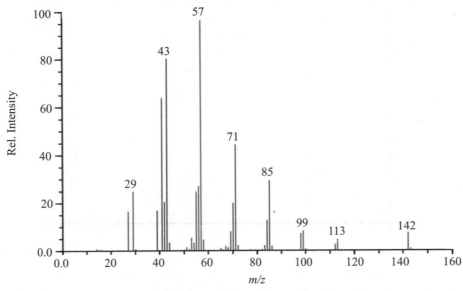

Sol. **(a)** A straight chain saturated hydrocarbon, **(b)** Molecular formula $C_{10}H_{22}$, **(c)** The molecular ion is separated by 15 mass units and other peaks in the spectrum differ by 14 mass units.

Q12. The mass spectrum of a compound known to be either 3- or 4-heptanone is shown below. Which ketone is most consistent with the data'?

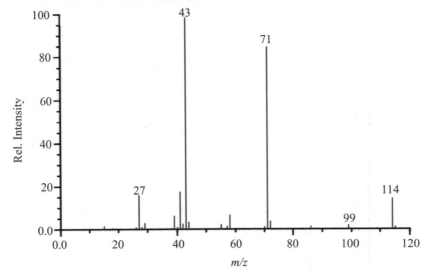

Sol. 4-heptanone

$$C_3H_7-\overset{\overset{\displaystyle O^{\bullet+}}{\|}}{C}-C_3H_7 \xrightarrow{-C_3H_7^{\bullet}} C_3H_7-C\equiv O^+$$

$$m/z\ 71$$

Q13. Suggest a structure for the molecule which gives the following mass spectrum.

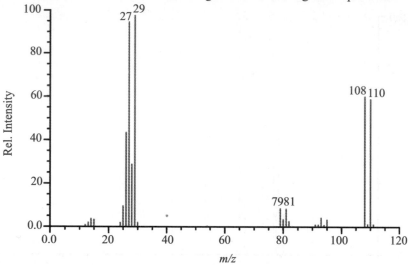

Sol. The isotope pattern of 1 : 1 of M and M+2 peaks suggest a mono-bromo compound. The probable formula is C_2H_5Br (M = 108 & M+2 = 110). m/z 79 and 81 are due to $(Br^{79})^+$ and $(Br^{81})^+$ formed by the homolytic hetero atom cleavage.

$$\left[C_2H_5\overset{\frown}{-}\overset{\bullet+}{Br}\right] \longrightarrow C_2H_5^+ + Br^{\bullet}$$

m/z 29 $(C_2H_5^+)$ is formed by the heterolytic heteroatom cleavage.

$$\left[C_2H_5\overset{\frown\frown}{-}\overset{\bullet+}{Br}\right] \longrightarrow C_2H_5^{\bullet} + Br^+$$

Q14. Suggest a structure for the molecule which gives the following mass spectrum.

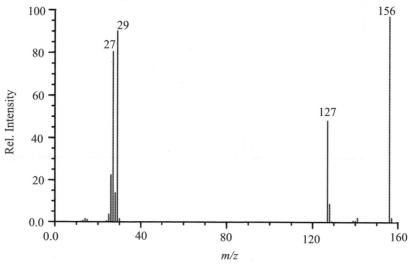

Sol. m/z 156 corresponds to $C_2H_5I^+$

m/z 127 is due to I^+ formed by the C-I cleavage.

$$\left[C_2H_5\overset{\frown}{\underset{}{\frown}}\overset{\bullet+}{Br} \right] \longrightarrow C_2H_5^\bullet + \overset{+}{Br}$$

m/z 29 ($C_2H_5^+$) is formed by the heterolytic heteroatom cleavage.

$$\left[C_2H_5\overset{\frown}{\underset{}{}}\overset{\bullet+}{I} \right] \longrightarrow C_2H_5^+ + I^\bullet$$
$$ m/z\ 127$$

Q15. Interpret the following mass specturum:

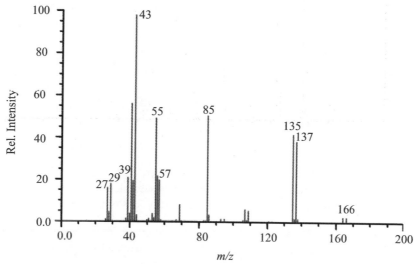

Sol. Loss of Br^+ results in generation of fragment at $m/z = 85$ and subsequent loss of 14 units results in other smaller fragment.

Q16. The mass spectrum of a compound known to be either methyl pentanoate or ethyl butanoate is shown below. Which structure is the most consistent with the data?

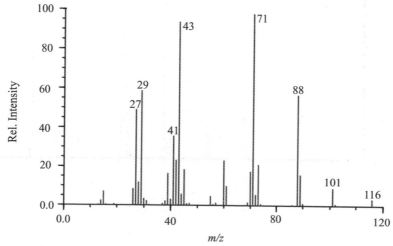

Sol. Ethyl butanoate

α-clevage products are $C_3H_7C{\equiv}O^+$ ($m/z = 71$); $C_3H_7^+$ ($m/z = 43$); $C_2H_5^+$ ($m/z = 29$).

Q17. Interpret the following mass spectrum:

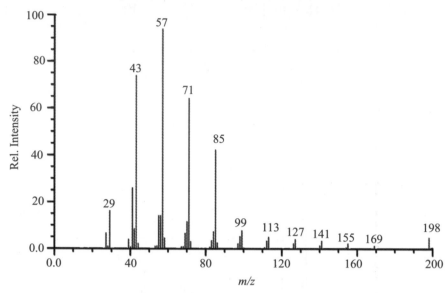

Sol. The molecular weight is consistent with the saturated hydrocarbon of molecular formula $C_{14}H_{30}$, the pattern of the peaks suggest an unbranched saturated structure. In the fragmentation pattern consecutive appearance of 14 units is due to the loss of CH_2 unit.

Q18. In the mass spectrum of $PhOCH_3$ an ion is observed at $m/z = 78$. What is the probable formula of the ion at $m/z = 78$? What neutral species has been extruded to form this ion?

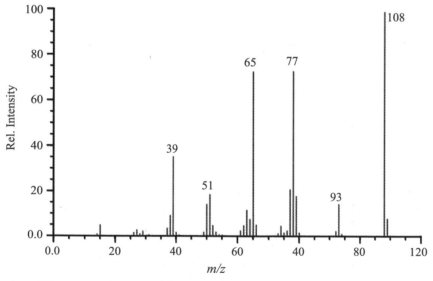

Sol. $C_6H_6^+$; CH_2O

UNSOLVED PROBLEMS

1. Decide: True or False?

   (a) Isotopic patterns depend on the elemental composition of an ion.

   (b) Isotopic patterns depend on the ionisation method used.

   (c) The monoisotopic peak is always the most abundant peak within an isotopic pattern.

   (d) The monoisotopic peak is always the peak of lowest m/z within an isotopic pattern.

   (e) Isotopic distributions depend on the charge state of an ion.

   (f) The charge state of an ion can be determined from the distance of isotopic peaks.

2. What does m/z mean?

3. What is an isotope?

4. Which are the common ways to normalise isotopic abundance?

5. Why do several signals appear in the mass spectrum?

6. How are relative abundance calculated in the mass spectrum?

7. Suggest a structure for the liquid compound, which gave a mass spectrum showing a strong molecular ion at $m/z = 156$. The only fragment ions are seen at $m/z = 127$ and 29.

8. Suggest a structure for the liquid compound gave a mass spectrum in which the molecular ion appears as a pair of equal intensity peaks at $m/z = 122$ and $m/z = 124$. Small fragment ion peaks are seen at $m/z = 107$ and 109 (equal intensity), and at $m/z = 79, 80, 81$, and 82 (all roughly the same size). Large fragment ions are seen at $m/z = 43$ (base peak), 41 and 39.

9. Calculate the accurate mass of the following ions:

   (a) H   (b) H_2O (c) C_6H_{12}

10. How are the ionizing electrons generated in electron ionisation (EI)?

11. Do EI spectra in general exhibit a high or a low degree of ion fragmentation?

12. Which of the following attributes apply to CI?

   (a) Soft ionisation technique   (b) Old method

   (c) Very difficult to apply   (d) Still highly relevant method

13. Consider a radical containing no nitrogen to be cleaved off from an even-mass molecular ion. Will the mass of the fragment be even- or odd-numbered?

14. Consider a molecule containing one nitrogen to be eliminated from a molecular ion containing two nitrogens. Will the mass of the fragment be even or odd-numbered?

15. Write down the α-cleavage of butanone molecular ion.

16. Show the structure of the ion that is responsible for the peak at $m/z = 43$ in the mass spectrum of 2-heptanone.

$$CH_3\overset{\overset{\displaystyle O}{\|}}{C}CH_2CH_2CH_2CH_2CH_3$$

17. Show the structure of the ion that is responsible for the peak at $m/z = 99$ in the mass spectrum of 2-heptanone.

18. Explain, how these isomers could be distinguished by mass spectroscopy.

19. McLafferty Rearrangement:

(i)

$$\xrightarrow{\text{McL}}$$

(ii)

$$\xrightarrow{\text{McL}}$$

(iii)

$$\xrightarrow{\text{McL}}$$

(iv)

$$\xrightarrow{\text{McL}}$$

20. Describe the applications of mass spectrometry.

21. Taking suitable example, explain the importance of McLafferty rearrangement on the mass spectra of organic compounds.

22. What are metastable ions? What is their significance?

23. Predict the mass spectrum of *n*-butyl acetate.

24. A ketone $C_7H_{14}O$ shows the following ion peaks in the mass spectrum. Deduce the structure and explain the genesis of ions, $m/z = 114, 85, 72, 57$ and 29.

25. Explain the following terms: (a) McLafferty rearrangement, (b) Molecular ion peak.

26. Assign the structure of a compound $C_{10}H_{12}O$, whose mass spectrum shows m/z values of 15, $43, 65, 57, 91, 105$ and 148.

27. Explain, how the peaks at $m/z = 115, 101$ and 73 arise in the mass spectrum of 3-methyl-3-heptanol.

28. The mass spectrum of 3-butyn-2-ol shows a large peak at $m/z = 55$. Draw the structure of the fragment and explain why it is particularly stable.

29. How can you use the information of ortho effect to explain the formation of the ion peak at $m/z = 149$ in diethyl phthalate?

30. Ethyl butanoate in its mass spectrum exhibits two characteristic peaks due to odd-electron ions at $m/z = 88$ and 60, and an abundant ion at $m/z = 70$. Explain the fragmentation pattern.

31. Indicate the major fragments that would be obtained in the MS of the following compounds:

    (a) Acetophenone   (b) Ethyl benzene

    (c) $CH_3CH_2CH_2COOCH_3$ (d) $C_6H_5CO(CH_2)_3CH_3$

32. Identify the bonds to be cleaved by RDA reaction of the following molecular ions.

(i)

$\xrightarrow{\text{RDA}}$

(ii)

$\xrightarrow{\text{RDA}}$

(iii)

$\xrightarrow{\text{RDA}}$

33. Why do RSH and RSR have small but significant M+2 peaks?

34. Why does the cyclohexane have an intense peak at $m/z = 54$?

35. The peak with the second highest m/z values in the spectrum of the compound $C_8H_{14}O_2$ is 114. List the compounds that could have been lost on fragmentation of the parent to give this peak.

36. Explain the appearance of $m/z = 44$ in the mass spectrum of $CH_3CH_2CH_2CHO$.

37. After the ionisation and fragmentation occurs what does the mass spectrometer do to provide a mass spectrum.

38. Why can the m/z values be assigned as the molecular weight of the cation in most cases?

39. How can the molecular weight of a compound be determined by mass spectral analysis?

40. How will you distinguish between the three isomeric butanols on the basis of mass spectrometry: 1-butanol, 2-butanol, 2-methyl-2-propanol.

41. How will you distinguish between the following amines on the basis of mass spectrometry: ethylamine, diethylamine, triethylamine.

42. How will you distinguish between o- and m-anisidine on the basis of mass spectrometry?

43. Alkylanilines exhibit the base peak at m/z 106. Name the ion which is responsible for it and explain.

44. Write the formula of the compound which is responsible for the appearance of a strong characteristic peak at m/z 149 in the mass spectra of all esters of phthalic acid. Describe the mechanism of its formation starting from diethyl phthalate.

45. How will you account for the appearance of prominent peaks at m/z 31, 42 and 70 in mass spectrum of n-pentanol.

46. How will you distinguish between 3-methyl and 4-methylcyclohexene on the basis of mass spectrometry?

47. Discuss the basic principles of mass spectrometry.

48. How can you distinguish between 2-pentanone from 3-pentanone by mass spectrometry?

49. A compound has molecular formula C_4H_7N; IR ($CHCl_3$): 2941, 2273, 1460 cm^{-1}; ^1H NMR ($CDCl_3$): $\delta = 2.72$ (sextet, 1H), 1.33 (d, 6H). Deduce the structure of the compound.

50. The mass spectrum of a compound known to be either 3- or 4-heptanone is shown below. Which ketone is most consistent with the data?

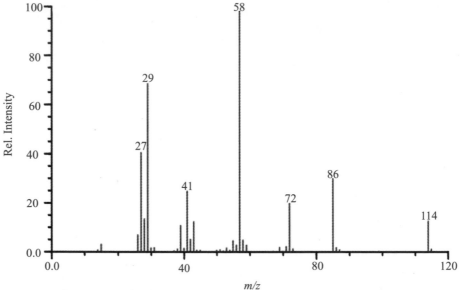

[**Hint:** 3-Heptanone]

51. Interpret the following mass spectrum:

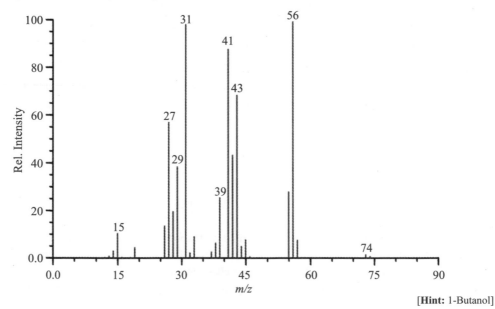

[**Hint:** 1-Butanol]

52. Account for the major peaks of each of the following compounds. Write the equation for their formation.

(a) $CH_3OCH_2CH_2CH_3$ ($m/z = 31, 45, 59$ and 74)

(b) *trans*-$CH_3CH=CHCH_2CH_2CH_3$ ($m/z = 41, 55$ and 84).

(c) $CH_3CH_2NHCH_2CH_2CH_3$ ($m/z = 30, 44, 58, 72$ and 87).

(d) $CH_3CH_2COOCH_3$ ($m/z = 43, 91, 147$ and 162).

53. In the mass spectrum of 2-pentanol no molecular ion is observed at $m/z = 88$. Assign a formula to the ion of $m/z = 73$.

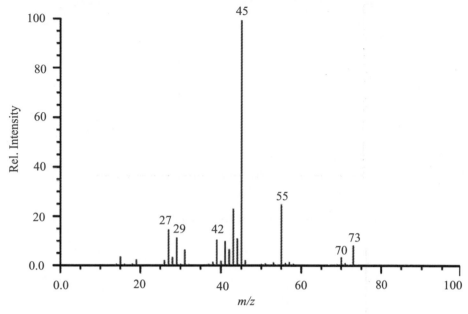

54. Interpret the following mass spectrum:

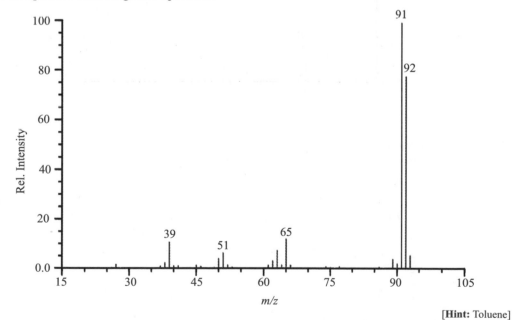

[**Hint:** Toluene]

55. The mass spectrum of a ketone is shown below, identify the structure of the compound.

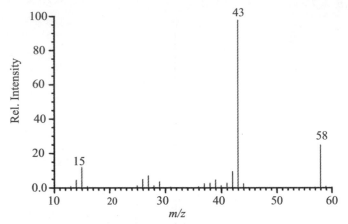

[**Hint:** Acetone]

56. Interpret the following mass spectrum of an alkyl halide:

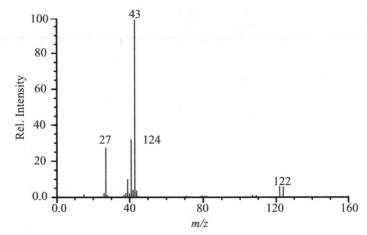

[**Hint:** Propyl bromide]

57. The mass spectrum of an aromatic ketone is shown below. Identify the compound and account for the formation and abundance of *m/z* 105.

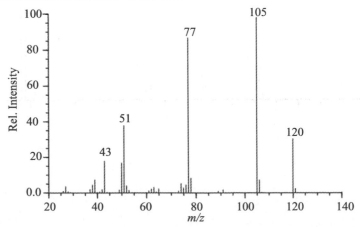

[**Hint:** Acetophenone]

58. Interpret the following mass spectrum:

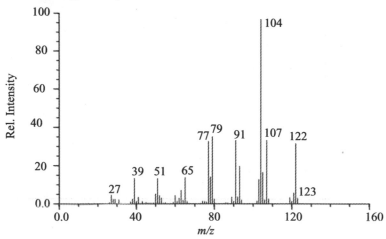

[**Hint:** 2-Methyl benzyl alcohol]

59. Interpret the following mass spectrum:

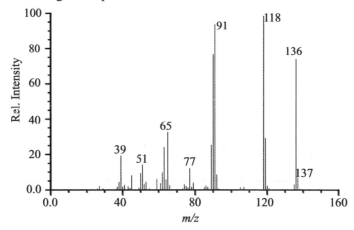

[**Hint:** 2-Methyl benzoic acid]

60. Interpret the following mass spectrum:

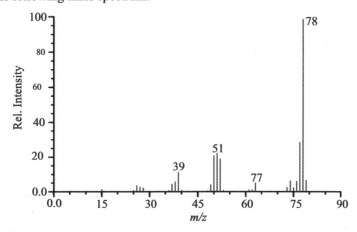

[**Hint:** Benzene]

61. Interpret the following mass spectrum:

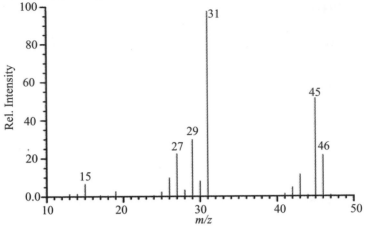

[**Hint:** Ethanol]

62. Interpret the following mass spectrum:

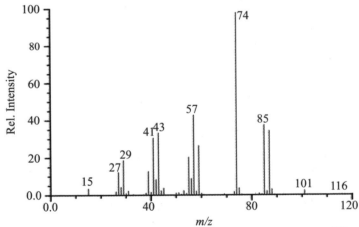

[**Hint:** Methyl pentanoate]

63. Interpret the following mass spectrum:

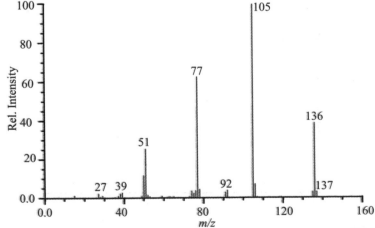

[**Hint:** Methyl benzoate]

64. Interpret the following mass spectrum:

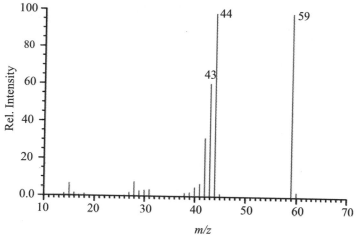

[**Hint:** Acetamide]

65. Suggest a structure to an amine by utilizing the following mass spectrum:

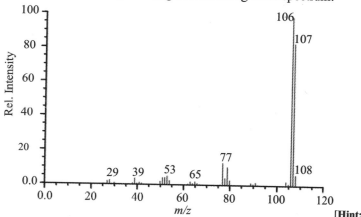

[**Hint:** *p*-Aminotoluene]

66. Suggest a structure to the hydrocarbon using the following mass spectrum.

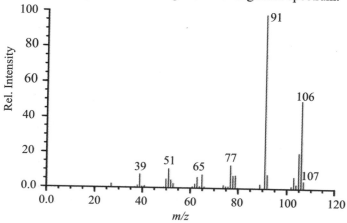

[**Hint:** o-Xylene]

Practice Problems

This chapter contains the questionnaire based on the techniques studied in the previous chapters such as UV-Visible, IR, NMR and Mass. The following step wise approach can be utilized for the interpretation of the spectral data:

Step I: A cursory examination of all the data to note the presence or absence of various structural features such as OH, C=O, aromatic protons, sites of unsaturation, etc.

Step II: A detailed examination of the spectra to determine the kinds of structural features present. (The NMR spectrum is perhaps the best to start with since it allows not only the determination of structural fragments but also calculation of possible molecular formulas if one is not available).

Step III: Prepare a list of all the structural fragments and molecular formulas determined. Elimination or combination of structural fragments by the restrictions of the molecular formula or, conversely, elimination of possible molecular formulas through the requirements of the structural fragments may be possible.

Step IV: Determine the residual formula and then combine fragments in all possible ways.

Step V: Examine the possible structure derived in step IV to see if any can be eliminated on the basis of other data that might be available.

1. Identify the structure of an organic compound by analysing the data given below, the molecular formula is $C_2H_3Cl_3$; IR (nujol, cm^{-1}): 2983, 1425, 1257, 1103, 841, 816.

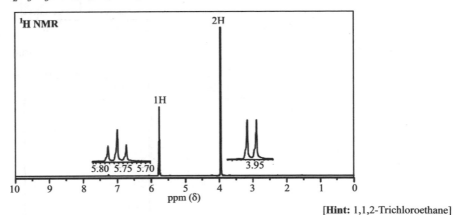

[**Hint:** 1,1,2-Trichloroethane]

2. Identify the structure of an organic compound by analysing the data given below, the molecular formula is $C_6H_{12}Br_2$; IR (nujol, cm^{-1}): 3011, 2950, 2885, 3019, 1415, 1254, 755, 825.

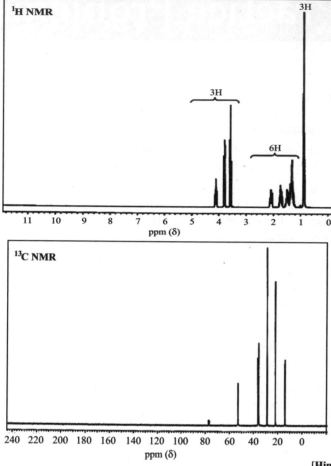

[**Hint:** 1,2-Dibromohexane]

3. Identify the structure of an organic compound by analysing the data given below, the molecular formula is $C_7H_{14}O$; IR (nujol, cm^{-1}): 2980, 2875, 1714, 1427, 1077.

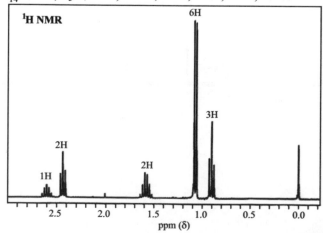

[**Hint:** 2-Methylhexan-3-one]

4. Identify the structure of an organic compound by analysing the data given below with molecular formula is $C_4H_8O_2$; IR (nujol, cm^{-1}): 3462, 2983, 2940, 2908, 2877, 1743, 1480, 1250, 1055.

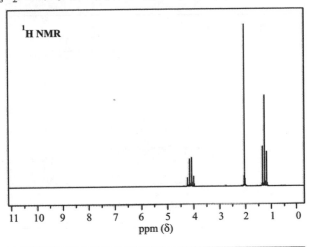

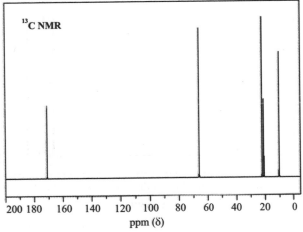

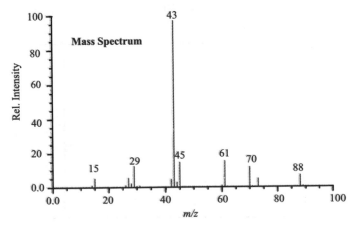

[Hint: Ethyl acetate]

5. Identify the structure of an organic compound by analysing the data given below, the molecular formula is $C_7H_{16}O$; IR (nujol, cm^{-1}): 3638, 3344, 2958, 2931, 1468, 1460, 1379, 1238, 1054, 1012.

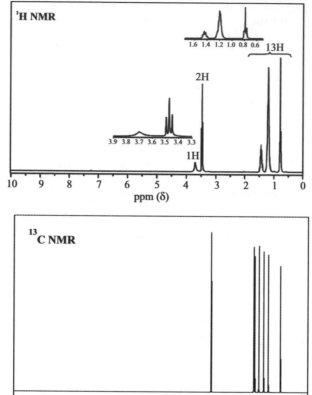

[**Hint:** 1-Heptanol]

6. Identify the structure of an organic compound by analysing the data given below, the molecular formula is C_3H_6O; IR (nujol, cm^{-1}): 3414, 3003, 2966, 1749, 1714, 1434, 1363, 1223, 1097, ^{1}H NMR (δ ppm): 2.01(6H).

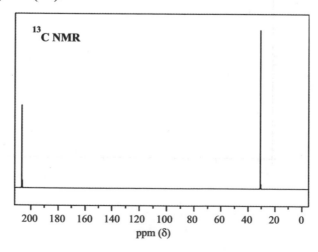

[**Hint:** Acetone]

7. Identify the structure of an organic compound by analysing the data given below, the molecular formula is C_2H_5I; IR (nujol, cm^{-1}): 3017, 2977, 2919, 2864, 1563, 1548, 1453, 1377, 1204, 952.

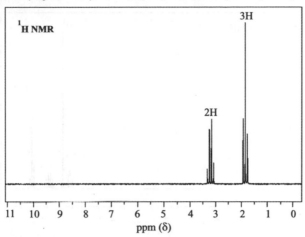

[**Hint:** Ethyl iodide]

8. Identify the structure of an organic compound by analysing the data given below, the molecular formula is C_7H_8; IR (nujol, cm^{-1}): 3099, 3032, 2925, 3019, 1614, 1506, 1465, 1086, 1035.

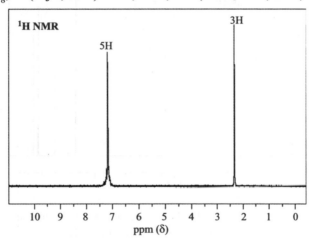

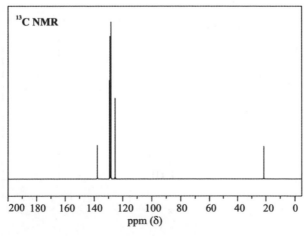

[**Hint:** Toluene]

9. Identify the structure by analysing the data given below, the molecular formula is $C_5H_{10}O$; IR (nujol, cm^{-1}): 3638, 3413, 2979, 1716, 1461, 1377, 1252, 1096, 1013.

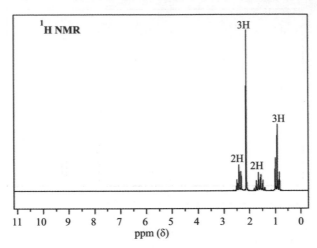

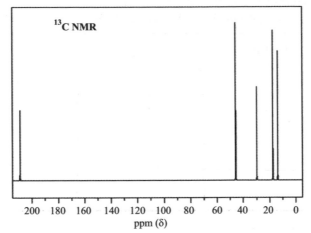

[**Hint:** 2-Pentanone]

10. Identify the structure by analysing the data given below, the molecular formula is $C_9H_8O_2$; IR (nujol, cm^{-1}): 2966, 2924, 1754, 1682, 1575, 1468, 1257, 1097.

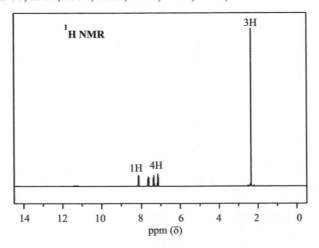

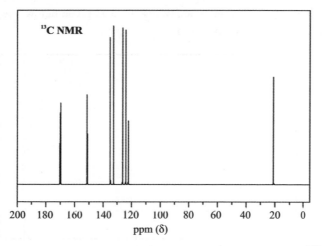

[**Hint:** Acetyl salicylic acid]

11. Identify the structure of an organic compound by analysing the data given below, the molecular formula is C_7H_8O; IR (nujol, cm^{-1}): 3501, 3450, 3319, 3019, 1704, 1257, 1177.

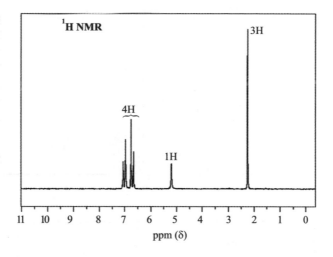

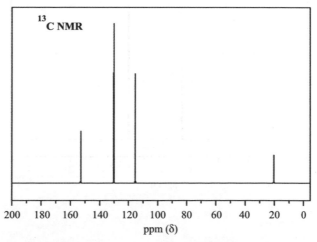

[**Hint:** *p*-Cresol]

12. Identify the structure of an organic by analysing the data given below, the molecular formula of compound is $C_2H_3Cl_3$; IR (nujol, cm^{-1}):

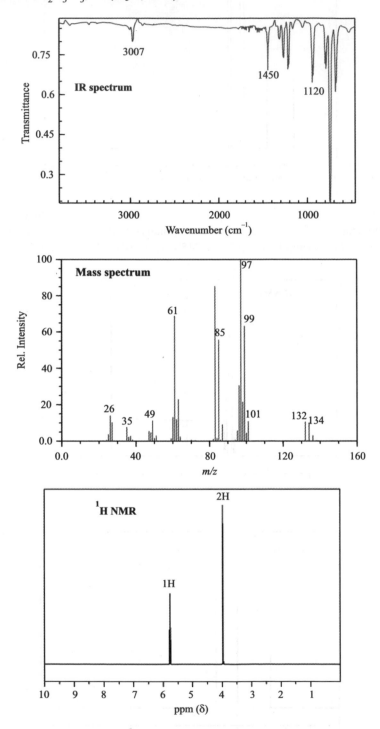

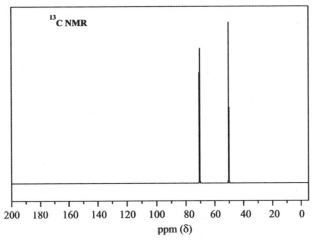

[Hint: 1,1,2-Trichloroethane]

13. Determine the structure of an organic compound by analysing the data given below, the molecular formula is $C_2H_{12}Br_2$; IR (nujol, cm^{-1}): 3011, 2950, 2885, 3019, 1415, 1254, 755, 825.

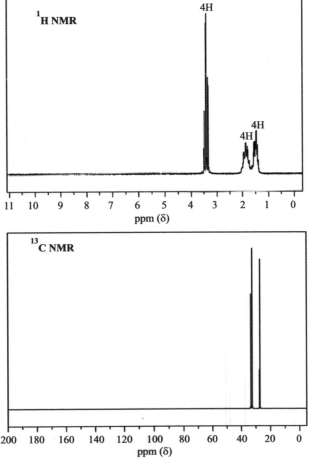

[Hint: 1,6-dibromohexane]

14. Identify the structure of an organic compound by analysing the data given below, the molecular formula is $C_7H_{14}O$; IR (nujol, cm^{-1}): 2980, 2875, 1714, 1427, 1077.

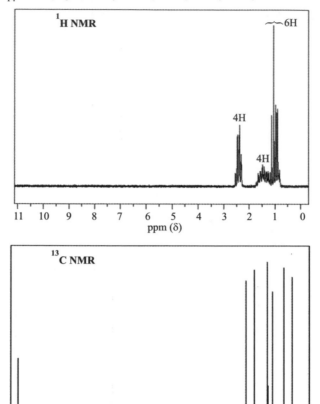

[**Hint:** 3-Heptanone]

15. Identify the structure of an organic compound by analysing the data given below, the molecular formula is $C_6H_5NO_2$.

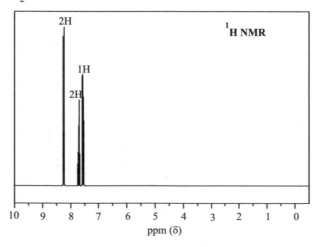

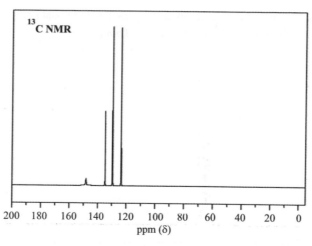

[**Hint:** Nitrobenzene]

16. Identify the structure of an organic compound by analysing the data given below, the molecular formula is $C_{10}H_{12}O$; IR (nujol, cm^{-1}): 3087, 3004, 2926, 2947, 1717, 1606, 1497, 1360, 750, 715.

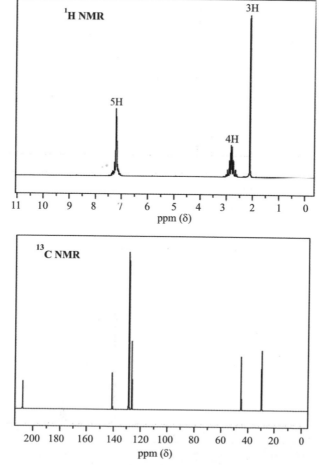

[**Hint:** 4-Phenyl-2-butanone]

17. Identify the structure of an organic compound by analysing the data given below, the molecular formula is $C_{14}H_{12}$; IR (nujol, cm^{-1}): 3077, 2966, 1660, 1452.

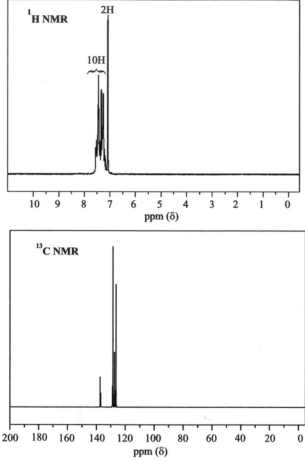

[Hint: *trans*-Stilbene]

18. Identify the structure of an organic compound by analysing the data given below, the molecular formula is $C_3H_8O_2$; IR (nujol, cm^{-1}): 3501, 3450, 1250, 1177.

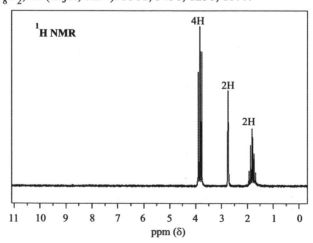

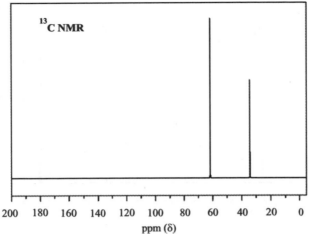

$$^{13}\text{C NMR}$$

200 180 160 140 120 100 80 60 40 20 0
ppm (δ)   [**Hint:** [1,3-Propanediol]]

19. Identify the structure of an organic compound by analysing the data given below, the molecular formula is $C_{11}H_{14}O_2$; IR (nujol, cm^{-1}): 1728, 1660, 1070.

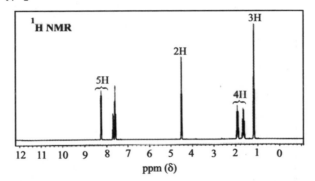

^{1}H NMR

12 11 10 9 8 7 6 5 4 3 2 1 0
ppm (δ)   [**Hint:** Butyl benzoate]

20. Identify the structure of an organic compound by analysing the data given below, the molecular formula is $C_5H_{10}O_2$; IR (nujol, cm^{-1}): 1743, 1243.

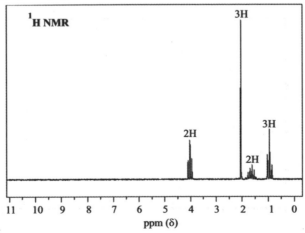

^{1}H NMR

11 10 9 8 7 6 5 4 3 2 1 0
ppm (δ)

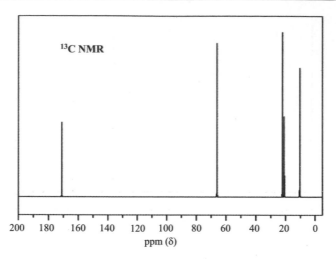

[**Hint:** Propyl acetate]

21. Identify the structure of compound with molecular formula C_3H_3N:

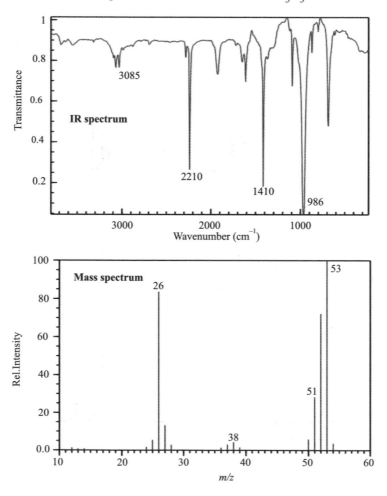

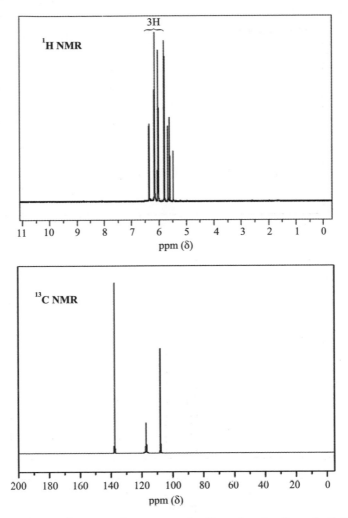

[**Hint:** Acrylonitrile]

22. Identify the structure of an organic compound with molecular formula $C_9H_8O_2$ by analysing the data given below.

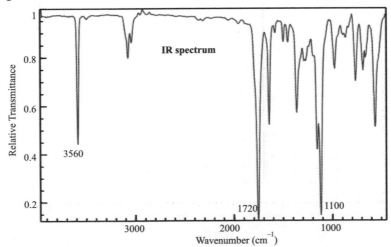

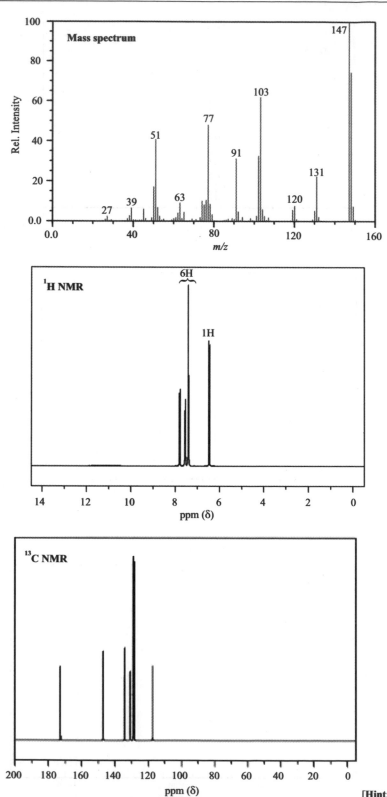

[**Hint:** Cinnamic acid]

23. Identify the structure of an organic compound with molecular formula $C_5H_4O_2$ by analysing the data given below.

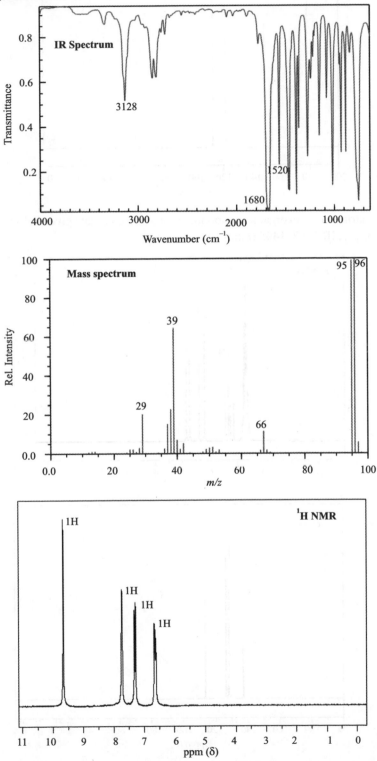

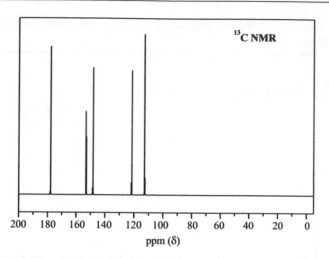

ppm (δ)

24. Identify the structure of an organic compound by analysing the data given below. The molecular formula is C_8H_8; IR 1162, 1445 (nujol cm^{-1}):

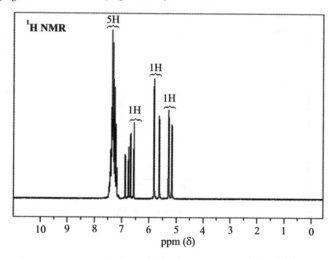

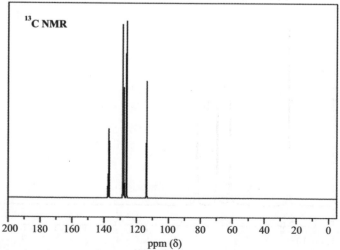

ppm (δ)

25. Identify the structure of an organic compound with molecular formula $C_8H_8O_2$ by analysing the data given below.

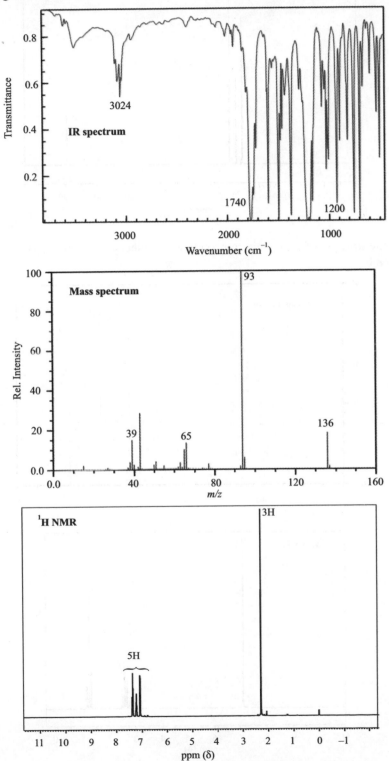

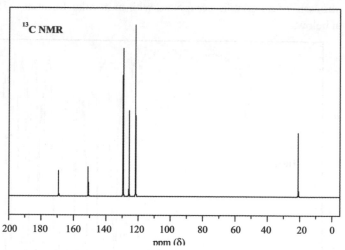

[**Hint:** Phenylacetate]

26. Identify the structure of an organic compound with molecular formula $C_{10}H_{14}$ by analysing the data given below.

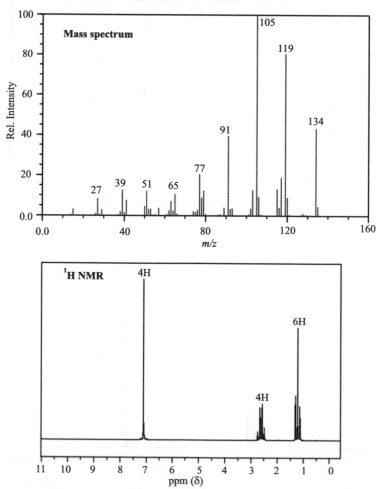

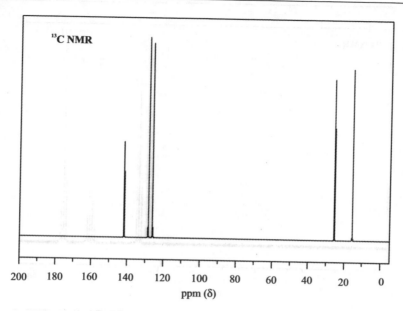

[Hint: 1,2-Diethylbenzene]

27. Identify the structure of an organic compound with molecular formula C_3H_7Br by analysing the data given below.

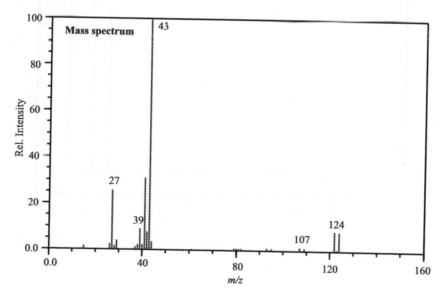

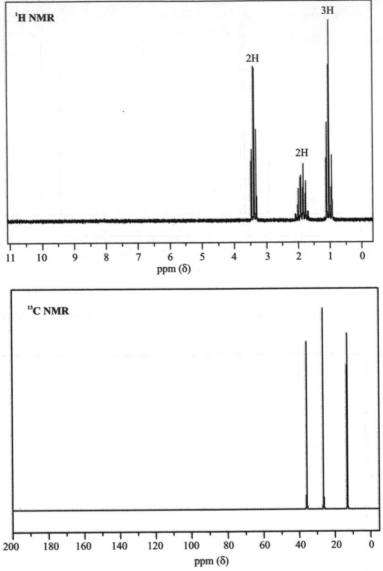

[**Hint:** 1-Bromopropane]

Appendix

Appendix 1

Fragments observed for Acids.

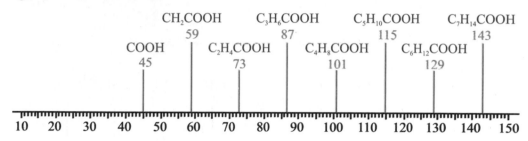

Appendix 2

Fragments observed for Aldehydes.

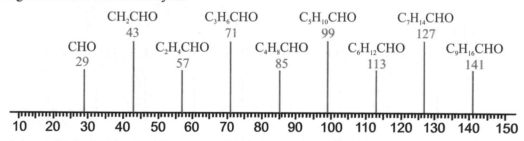

Appendix 3

Fragments observed for Alkanes.

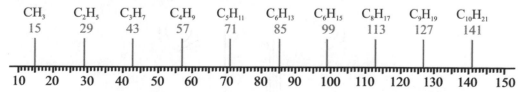

Appendix 4

Fragments observed for Alkenes.

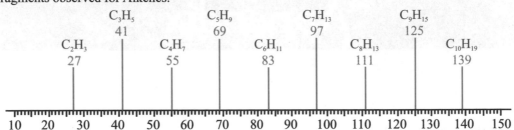

Appendix 5

Fragments observed for Alkynes.

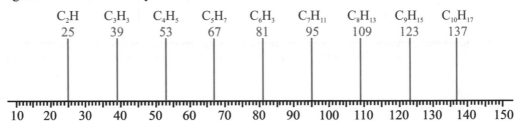

Appendix 6

Fragments observed for Amines.

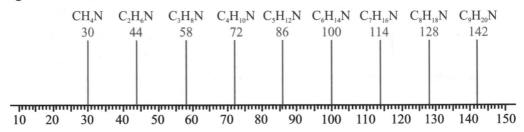

Appendix 7

Fragments observed for Aromatics.

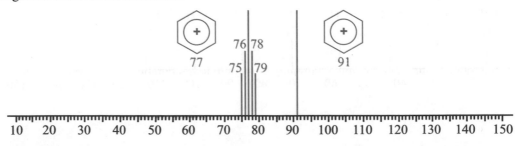

Appendix 8

Fragments observed for Esters.

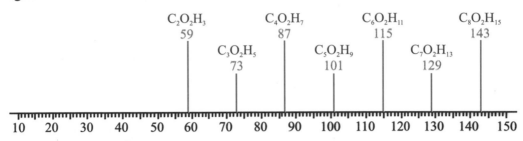

Appendix 9

Fragments observed for Alcohols, Ethers.

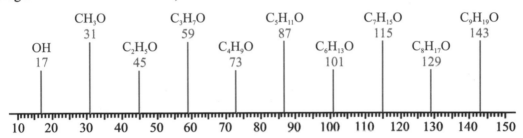

Appendix 10

Fragments observed for Ketones.

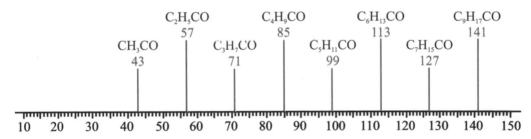

Index